T0214024

Lecture Notes in Mathematics

Volume 2297

Editors-in-Chief

Jean-Michel Morel, CMLA, ENS, Cachan, France

Bernard Teissier, IMJ-PRG, Paris, France

Series Editors

Karin Baur, University of Leeds, Leeds, UK

Michel Brion, UGA, Grenoble, France

Alessio Figalli, ETH Zurich, Zurich, Switzerland

Annette Huber, Albert Ludwig University, Freiburg, Germany

Davar Khoshnevisan, The University of Utah, Salt Lake City, UT, USA

Ioannis Kontoyiannis, University of Cambridge, Cambridge, UK

Angela Kunoth, University of Cologne, Cologne, Germany

László Székelyhidi, Institute of Mathematics, Leipzig University, Leipzig, Germany

Ariane Mézard, IMJ-PRG, Paris, France

Mark Podolskij, University of Luxembourg, Esch-sur-Alzette, Luxembourg

Sylvia Serfaty, NYU Courant, New York, NY, USA

Gabriele Vezzosi, UniFI, Florence, Italy

Anna Wienhard, Ruprecht Karl University, Heidelberg, Germany

This series reports on new developments in all areas of mathematics and their applications - quickly, informally and at a high level. Mathematical texts analysing new developments in modelling and numerical simulation are welcome. The type of material considered for publication includes:

1. Research monographs
2. Lectures on a new field or presentations of a new angle in a classical field
3. Summer schools and intensive courses on topics of current research.

Texts which are out of print but still in demand may also be considered if they fall within these categories. The timeliness of a manuscript is sometimes more important than its form, which may be preliminary or tentative.

Titles from this series are indexed by Scopus, Web of Science, Mathematical Reviews, and zbMATH.

More information about this series at http://www.springer.com/series/304

Valentina Georgoulas • Joel W. Robbin •
Dietmar Arno Salamon

The Moment-Weight Inequality and the Hilbert–Mumford Criterion

GIT from the Differential Geometric Viewpoint

Springer

Valentina Georgoulas
Zürich, Switzerland

Joel W. Robbin
Department of Mathematics
University of Wisconsin–Madison
Madison, WI, USA

Dietmar Arno Salamon
Department of Mathematics
ETH Zürich
Zürich, Switzerland

ISSN 0075-8434 ISSN 1617-9692 (electronic)
Lecture Notes in Mathematics
ISBN 978-3-030-89299-9 ISBN 978-3-030-89300-2 (eBook)
https://doi.org/10.1007/978-3-030-89300-2

Mathematics Subject Classification: 53D20, 14L24, 32Q15

© The Editor(s) (if applicable) and The Author(s), under exclusive license to Springer Nature Switzerland AG 2021
This work is subject to copyright. All rights are solely and exclusively licensed by the Publisher, whether the whole or part of the material is concerned, specifically the rights of translation, reprinting, reuse of illustrations, recitation, broadcasting, reproduction on microfilms or in any other physical way, and transmission or information storage and retrieval, electronic adaptation, computer software, or by similar or dissimilar methodology now known or hereafter developed.
The use of general descriptive names, registered names, trademarks, service marks, etc. in this publication does not imply, even in the absence of a specific statement, that such names are exempt from the relevant protective laws and regulations and therefore free for general use.
The publisher, the authors, and the editors are safe to assume that the advice and information in this book are believed to be true and accurate at the date of publication. Neither the publisher nor the authors or the editors give a warranty, expressed or implied, with respect to the material contained herein or for any errors or omissions that may have been made. The publisher remains neutral with regard to jurisdictional claims in published maps and institutional affiliations.

This Springer imprint is published by the registered company Springer Nature Switzerland AG.
The registered company address is: Gewerbestrasse 11, 6330 Cham, Switzerland

Preface

This book gives an essentially self-contained exposition (except for an appeal to the Lojasiewicz gradient inequality) of geometric invariant theory from a differential geometric viewpoint. Central ingredients are the moment-weight inequality (relating the Mumford numerical invariants to the norm of the moment map), the negative gradient flow of the moment map squared, and the Kempf–Ness function.

The last author, DAS, owes a lot to the lectures by and conversations with Simon Donaldson, Gábor Székelyhidi, Xiuxiong Chen, and Song Sun. The second author, JWR, learned much from a course given by and conversations with Sean Paul at the University of Wisconsin. Most of this book was written when JWR visited the Forschungsinstitut für Mathematik at ETH Zürich and he thanks them for their hospitality. The first version of this book was completed while DAS visited the IAS, Princeton, and the SCGP, Stony Brook; he thanks both institutes for their hospitality. Thanks to Samuel Trautwein for helpful discussions. Thanks to Amanda Jenny for pointing out errors in earlier versions of the manuscript.

Zürich, Switzerland — Valentina Georgoulas
Madison, WI, USA — Joel W. Robbin
Zürich, Switzerland — Dietmar A. Salamon
August 2021

Contents

Chapter 1
Introduction

Many important problems in geometry can be reduced to a partial differential equation of the form

$$\mu(x) = 0,$$

where x ranges over a complexifed group orbit in an infinite-dimensional symplectic manifold X and $\mu : X \to \mathfrak{g}$ is an associated moment map (see Calabi [6–8], Yau [79–81], Tian [75], Chen–Donaldson–Sun [18–20], Atiyah–Bott [2], Uhlenbeck–Yau [76], Donaldson [27–30, 32]). Problems like this are extremely difficult. The purpose of this book is to explain the analogous finite-dimensional situation, which is the subject of Geometric Invariant Theory.

GIT was originally developed to study actions of a complex reductive Lie group G^c on a projective algebraic variety[1]

$$X \subset \mathbb{P}(V).$$

Here G^c is the complexification of a compact Lie group G and the Hermitian structure on V can be chosen so that G acts by unitary automorphisms. In the smooth case X inherits the structure of a Kähler manifold from the standard Kähler structure on $\mathbb{P}(V)$. The G-action is generated by the standard moment map

$$\mu : X \to \mathfrak{g}$$

(with values in the Lie algebra of G). In the original treatment of Mumford [63] the symplectic form and the moment map were not used. Subsequently several authors

[1] Mumford's original theory even applies to varieties over general ground fields other than the complex numbers. These extensions are not discussed in our treatment.

© The Author(s), under exclusive license to Springer Nature Switzerland AG 2021

V. Georgoulas et al., *The Moment-Weight Inequality and the Hilbert–Mumford Criterion*, Lecture Notes in Mathematics 2297,
https://doi.org/10.1007/978-3-030-89300-2_1

discovered the connection between the theory of the moment map and GIT (see Kirwan [53] and Ness [66]). In the article of Lerman [56] it is noted that both Guilleman–Sternberg [42] and Ness [66] credit Mumford for the relation between the complex quotient and the Marsden–Weinstein quotient

$$X/\!\!/\mathrm{G} := \mu^{-1}(0)/\mathrm{G}.$$

In our exposition we assume that X is a closed Kähler manifold but do not assume that it is a projective variety. In the latter case the aforementioned standard moment map satisfies certain rationality conditions (Chap. 9) which we do not use in our treatment. As a result the Mumford numerical invariants

$$w_\mu(x,\xi) := \lim_{t\to\infty} \langle \mu(\exp(\mathbf{i}t\xi)x), \xi\rangle$$

associated to a point $x \in X$ and an element $\xi \in \mathfrak{g}\setminus\{0\}$ (Chap. 5) need not be integers as they are in traditional GIT. In the classical theory the Lie algebra element belongs to the set

$$\Lambda := \{\xi \in \mathfrak{g}\setminus\{0\} \mid \exp(\xi) = \mathbb{1}\}$$

and thus generates a one-parameter subgroup of G^c. In our treatment ξ can be a general nonzero element of $\mathfrak{g}$.

A central ingredient in our treatment is the **moment-weight inequality**

$$\sup_{\xi\in\mathfrak{g}\setminus\{0\}} \frac{-w_\mu(x,\xi)}{|\xi|} \le \inf_{g\in\mathrm{G}^c} |\mu(gx)|. \tag{1.1}$$

We give two proofs of this inequality in Chap. 6, one due to Mumford [63] and Ness [66, Lemma 3.1] and one due to Xiuxiong Chen [15]. (For an in depth discussion of this inequality see Atiyah–Bott [2], in the setting of bundles over Riemann surfaces, and Donaldson [34], Székelyhidi [72], Chen [15, 16], in the setting of Kähler–Einstein geometry.) Following an argument of Chen–Sun [23] we also prove that equality holds in (1.1) whenever the right hand side is positive (Theorem 10.4). We also prove that the supremum on the left is always attained (Theorem 10.1) and that the supremum over all $\xi \in \mathfrak{g}\setminus\{0\}$ agrees with the supremum over all $\xi \in \Lambda$ (Theorem 12.1). In the projective case the supremum is attained at an element $\xi \in \Lambda$ by a theorem of Kempf [51], however, that need not be the case in our more general setting. The **Hilbert–Mumford numerical criterion** for μ-semistability is an immediate consequence of the aforementioned results (Theorem 12.2). It asserts that

$$\overline{\mathrm{G}^c(x)} \cap \mu^{-1}(0) \neq \emptyset \qquad \iff \qquad w_\mu(x,\xi) \ge 0 \;\forall \xi \in \Lambda.$$

Further consequences of the moment-weight inequality include the **Kirwan–Ness Inequality** which asserts that if x is a critical point of the moment map squared then

$$|\mu(x)| = \inf_{g \in \mathrm{G}^c} |\mu(gx)|$$

(Corollary 6.2), the **Moment Limit Theorem** which asserts that each negative gradient flow line of the moment map squared converges to a minimum of the moment map squared on the complexified group orbit (Theorem 6.4), and the **Ness Uniqueness Theorem** which asserts that any two critical points of the moment map squared in the same G^c-orbit in fact belong to the same G-orbit (Theorem 6.3) and, moreover, that the minimum of the moment map squared on the closure of a G^c-orbit is taken on at a unique G-orbit (Theorem 6.5).

A central ingredient in the proofs of these theorems is the negative gradient flow of the moment map squared

$$f := \tfrac{1}{2}|\mu|^2 : X \to \mathbb{R}.$$

The gradient flow equation takes the form

$$\dot{x} = -J L_x \mu(x) \tag{1.2}$$

where $L_x : \mathfrak{g} \to T_x X$ denotes the infinitesimal action of the Lie algebra on X. Each G^c-orbit $\mathrm{G}^c(x) \subset X$ is invariant under this flow, because every solution of (1.2) has the form $x(t) = g(t)^{-1}x$, where $g : \mathbb{R} \to \mathrm{G}^c$ satisfies the differential equation

$$g^{-1}\dot{g} = \mathbf{i}\mu(g^{-1}x). \tag{1.3}$$

Equation (1.3) is the negative gradient flow of the **Kempf–Ness function**

$$\Phi_x : \mathrm{G}^c/\mathrm{G} \to \mathbb{R}.$$

The homogeneous space G^c/G is simply connected and complete with nonpositive sectional curvature, and the Kempf–Ness function is Morse–Bott and is convex along geodesics (Theorem 4.3); its critical manifold may be empty. Moreover, the **Kempf–Ness Theorem** characterizes the stability conditions in terms of the properties of the Kempf–Ness function (Theorem 7.3); for example a point $x \in X$ is μ-semistable, i.e. the closure of its G^c-orbit intersects the zero set of the moment map, if and only if the Kempf–Ness function Φ_x is bounded below.

The moment map squared is in general far from being Morse–Bott and may have very complicated critical points. However, the aforementioned theorems (Kirwan–Ness Inequality, Ness Uniqueness, Moment Limit Theorem) exhibit a structure of the gradient flow that resembles the stratification by stable manifolds associated to a Morse–Bott function. More precisely, an element $x \in X$ is a critical point of the moment map squared if it satisfies the equation $L_x \mu(x) = 0$. Critical points come in G-orbits and the theorems of Ness and Kirwan–Ness show that the stable manifold of such a critical orbit $\mathrm{G}(x)$ is a union of G^c-orbits, i.e.

$$W^s(\mathrm{G}(x)) = \left\{ y \in X \,\middle|\, x \in \overline{\mathrm{G}^c(y)},\ |\mu(x)| = \inf_{\mathrm{G}^c(y)} |\mu| \right\}. \tag{1.4}$$

This stratification was used by Kirwan [53] to prove that the canonical ring homomorphism (called the **Kirwan homomorphism**)

$$\kappa : H^*_{\mathrm{G}}(X) \to H^*(X/\!\!/\mathrm{G})$$

from the equivariant cohomology of X to the cohomology of the Marsden–Weinstein quotient $X/\!\!/\mathrm{G}$ with rational coefficients is surjective. Kirwan's theorem is not included in our treatment. The main motivation for the present book comes from infinite-dimensional analogues of GIT in various areas of geometry.

One such infinite-dimensional analogue of geometric invariant theory is the Donaldson–Uhlenbeck–Yau correspondence between stable holomorphic vector bundles and Hermitian Yang–Mills connections over Kähler manifolds. (This is a special case of the Kobayashi–Hitchin correspondence.) In this theory the space of Hermitian connections on a Hermitian vector bundle over a closed Kähler manifold with curvature of type (1, 1) is viewed as an infinite-dimensional symplectic manifold, the group of unitary gauge transformations acts on it by Hamiltonian symplectomorphisms, the moment map assigns to a connection the component of the curvature parallel to the symplectic form, and the zero set of the moment map is the space of Hermitian Yang–Mills connections. For bundles over Riemann surfaces the analogue of the Hilbert–Mumford numerical criterion is the correspondence between stable bundles and flat connections, established by Narasimhan–Seshadri [64]. It can be viewed as an extension to higher rank bundles of the identification of the Jacobian with a torus. The moment map picture in this setting was exhibited by Atiyah–Bott [2] and they proved the analogue of the moment-weight inequality. Another proof of the Narasimhan–Seshadri theorem was given by Donaldson [27]. In dimension four the Hermitian Yang–Mills connections are anti-self-dual instantons over Kähler surfaces. In this setting the correspondence between stable bundles and ASD instantons was established by Donaldson [28] and used to prove nontriviality of the Donaldson invariants for Kähler surfaces. Donaldson's theorem was extended to higher-dimensional Kähler manifolds by Uhlenbeck–Yau [76].

Another infinite-dimensional analogue of GIT is Donaldson's program for the study of constant scalar curvature Kähler (cscK) metrics. It was noted by Donaldson [30] and Fujiki [39] that the scalar curvature can be interpreted as a moment map for the action of the group of Hamiltonian symplectomorphisms of a symplectic manifold (V, ω_0) on the space $\mathscr{J}_0$ of ω_0-compatible integrable complex structures. It was also noted by Donaldson [30] that in this setting the Futaki invariants [40] are analogues of the Mumford numerical invariants, the space of Kähler potentials is the analogue of G^c/G, the Mabuchi functional [58, 59] is the analogue of the log-norm function in Kempf–Ness [52], and that Tian's notion of K-stability [75] can be understood as an analogue of stability in GIT. This led to Donaldson's conjecture relating K-stability to the existence of cscK metrics [32, 33] and refining earlier conjectures by Yau [81] and Tian [75]. The Yau–Tian–Donaldson conjecture is the analogue of the Hilbert–Mumford criterion for μ-polystability (Theorem 12.5), with the μ-weights replaced by Donaldson's generalized Futaki invariants. The earlier conjecture of Yau applies to Fano manifolds, relates K-stability to the existence of Kähler–Einstein metrics, and was recently confirmed by Chen–Donaldson–Sun [17–20, 37]. The Yau–Tian–Donaldson conjecture, in the case where the first Chern class is not a multiple of the Kähler class, is still open. The moment-weight inequality in this setting was proved by Donaldson [34] and Chen [15, 16].

In this situation the duality between the positive curvature manifold G and the negative curvature manifold G^c/G is particularly interesting. The analogue of G is the group $\mathscr{G}_0$ of Hamiltonian symplectomorphisms of (V, ω_0) with the L^2 inner product on the Lie algebra of Hamiltonian functions, and the analogue of G^c/G is the space $\mathscr{H}_0$ of Kähler potentials on (V, J_0, ω_0) with an L^2 Riemannian metric. On the one hand the distance function associated to the L^2 metric on $\mathscr{G}_0$ is trivial by a result of Eliashberg–Polterovich [38]. On the other hand Chen [14] proved that in $\mathscr{H}_0$ any two points are joined by a unique geodesic of class $C^{1,1}$ (a solution of the Monge–Ampère equation), that $\mathscr{H}_0$ is a genuine metric space,[2] and that cscK metrics with negative first Chern class are unique in their Kähler class. The latter result was extended by Donaldson [31] to all cscK metrics with discrete automorphism groups. Calabi–Chen [9] proved that $\mathscr{H}_0$ is negatively curved in the sense of Alexandrov and, assuming all geodesics are smooth, that extremal metrics are unique up to holomorphic diffeomorphism. (Extremal metrics are the analogues of critical points of the moment map squared and their uniqueness up to holomorphic diffeomorphism is the analogue of the Ness Uniqueness Theorem 6.3.) Chen–Tian [25] removed the hypothesis on the geodesics. They also proved that cscK metrics minimize the Mabuchi functional. This was independently proved by Donaldson [35] in the projective case. As noted by Chen [16] and Chen–Sun [22], several straight forward statements in GIT have analogues in the cscK setting that are open questions. At the time one of these was convexity of the Mabuchi functional along $C^{1,1}$ geodesics. This has now been settled by Berman–Berndtsson [4] and Chen–Li–Paun [21]. In [67] S.T. Paul introduced another notion of stability and in [68] he proved that it is equivalent to a properness condition on certain finite-dimensional approximations of the Mabuchi functional. For Fano manifolds with discrete automorphism groups it has been shown that P-stability is equivalent to the aforementioned K-stability condition of Yau–Tian–Donaldson, by combining the work of Chen–Donaldson–Sun and S.T. Paul with a partial C^0 estimate of Gábor Székelyhidi [73] (see Chen–Sun–Wang [24]).

[2] The metric completion of the space $\mathscr{H}_0$ of Kähler potentials has been studied by Tamás Darvas [26].

The emphasis in the present book is on self contained proofs in the finite-dimensional setting. For other expositions of geometric invariant theory and the moment map see the papers by Thomas [74], which includes a detailed discussion of the cscK analogue, and by Woodward [78], which contains many finite-dimensional examples.

Here is a brief description of the content of the book. Two preliminary chapters introduce the basic setup and the moment map (Chap. 2) and examine the negative gradient flow of the moment map squared (Chap. 3). Chapter 4 introduces the Kempf–Ness function and establishes its basic properties. The Mumford numerical invariants are defined in Chap. 5 and shown to be invariant under the G^c-action and under the Mumford equivalence relation on the set of toral generators. Chapter 6 establishes the moment-weight inequality and derives several consequences such as the Kirwan–Ness Inequality, the Moment Limit Theorem, and the Ness Uniqueness Theorem. The μ-stability notions are introduced in Chap. 7 which also proves the Kempf–Ness Theorem. Chapter 8 deals with the classical algebraic geometric setting of linear actions on projective varieties by reductive groups and shows how it fits into the symplectic setup. Chapter 9 deals with the converse question and examines under which rationality conditions on the symplectic form and the moment map the general symplectic setting of the present book reduces to the classical algebro geometric setting. Chapter 10 is devoted to the Kempf Existence Theorem and shows that in the μ-unstable case the moment-weight inequality is actually an equality. It also shows that the supremum in (1.1) is always attained. That the supremum over $\mathfrak{g} \setminus \{0\}$ in (1.1) agrees with the supremum over Λ requires continuous dependence of the weight $w_\mu(x, \xi)$ on ξ for torus actions and this is the subject of Chap. 11. The Hilbert–Mumford criterion is proved in Chap. 12. Chapter 13 explains a criterion by Gabor Székelyhidi for points in X whose G^c-orbits contain critical points of the moment map squared. The criterion takes the form of polystability with respect to the action of a suitable subgroup. Several examples are discussed in Chap. 14.

Five appendices deal with relevant background material. Appendix A establishes some properties of complete simply connected Riemannian manifolds with non-positive sectional curvature and contains a proof of Cartan's Fixed Point Theorem. Appendix B establishes the existence of a complexification G^c of a compact Lie group G and shows how it is characterized by a universality property. Appendix C shows that the homogeneous space G^c/G is a complete simply connected Riemannian manifold with nonpositive sectional curvature. Appendix D introduces parabolic subgroups and the Mumford equivalence relation on the space of toral generators and Appendix E is devoted to the proof that each element of G^c factorizes as a product of an element of a given parabolic subgroup and an element of G.

Chapter 2
The Moment Map

Throughout (X, ω, J) denotes a closed Kähler manifold, i.e. X is a compact manifold without boundary, ω is a symplectic form on X, and J is an integrable complex structure on X, compatible with ω so

$$\langle \cdot, \cdot \rangle := \omega(\cdot, J\cdot)$$

is a Riemannian metric. Denote by ∇ the corresponding Levi-Civita connection.

The Moment Map Let $\mathrm{G} \subset \mathrm{U}(n)$ be a compact Lie group acting on X by Kähler isometries so that the action of G preserves all three structures $\langle \cdot, \cdot \rangle$, ω, J. Denote the group action by

$$\mathrm{G} \times X \to X : (u, x) \mapsto ux$$

and the infinitesimal action of the Lie algebra $\mathfrak{g} := \mathrm{Lie}(\mathrm{G}) \subset \mathfrak{u}(n)$ by

$$\mathfrak{g} \to \mathrm{Vect}(X) : \xi \mapsto v_\xi.$$

We assume that $\mathfrak{g}$ is equipped with an invariant inner product and that the group action is Hamiltonian. Let $\mu : X \to \mathfrak{g}$ be a moment map for the action, i.e. v_ξ is a Hamiltonian vector field with Hamiltonian function

$$H_\xi := \langle \mu, \xi \rangle$$

so $\iota(v_\xi)\omega = dH_\xi$ or, equivalently,

$$\langle d\mu(x)\widehat{x}, \xi \rangle = \omega(v_\xi(x), \widehat{x}) \tag{2.1}$$

© The Author(s), under exclusive license to Springer Nature Switzerland AG 2021

V. Georgoulas et al., *The Moment-Weight Inequality and the Hilbert–Mumford Criterion*, Lecture Notes in Mathematics 2297,
https://doi.org/10.1007/978-3-030-89300-2_2

for $x \in X, \widehat{x} \in T_x X$, and $\xi \in \mathfrak{g}$. We assume that the moment map is equivariant, i.e.

$$\mu(ux) = u\mu(x)u^{-1}, \qquad \langle \mu(x), [\xi, \eta] \rangle = \omega(v_\xi(x), v_\eta(x)) \tag{2.2}$$

for $x \in X$, $u \in \mathrm{G}$, and $\xi, \eta \in \mathfrak{g}$. The two equations in (2.2) are equivalent whenever G is connected.

The Complexified Group The map

$$\mathrm{U}(n) \times \mathfrak{u}(n) \to \mathrm{GL}(n, \mathbb{C}) : (u, \eta) \mapsto \exp(\mathbf{i}\eta)u$$

is a diffeomorphism, by polar decomposition, and the image of $\mathrm{G} \times \mathfrak{g}$ under this diffeomorphism is denoted by

$$\mathrm{G}^c := \{hu \mid u \in \mathrm{G},\ h = \exp(\mathbf{i}\eta),\ \eta \in \mathfrak{g}\}. \tag{2.3}$$

This is a Lie subgroup of $\mathrm{GL}(n, \mathbb{C})$,[1] called the **complexification of** G (see Theorem B.4). It contains G as a maximal compact subgroup, the quotient G^c/G is connected, and the Lie algebra of G^c is the complexification

$$\mathfrak{g}^c := \mathrm{Lie}(\mathrm{G}^c) = \mathfrak{g} \oplus \mathbf{i}\mathfrak{g}$$

of the Lie algebra of G (see Theorem B.2). We will consistently use the notations

$$\zeta = \xi + \mathbf{i}\eta, \qquad \mathrm{Re}(\zeta) := \xi, \qquad \mathrm{Im}(\zeta) := \eta.$$

for the elements $\zeta \in \mathfrak{g}^c$ and their real and imaginary parts $\xi, \eta \in \mathfrak{g}$. The reader is cautioned that the eigenvalues of ξ, η are imaginary and the eigenvalues of $\mathbf{i}\xi, \mathbf{i}\eta$ are real. A complex Lie group is called **reductive** iff it is the complexification of a compact Lie group.

The Action of the Complexified Group Let $\mathrm{G}^c \subset \mathrm{GL}(n, \mathbb{C})$ be the complexification of G and let $\mathfrak{g}^c = \mathfrak{g} + \mathbf{i}\mathfrak{g}$ be its Lie algebra. Then every Lie group homomorphism from G to a complex Lie group extends uniquely to a homomorphism from G^c to that complex Lie group (see Theorem B.2). Taking the target group to be the group of holomorphic automorphisms of X one obtains a holomorphic group action of G^c on X. Denote the group action by

$$\mathrm{G}^c \times X \to X : (g, x) \mapsto gx$$

and the infinitesimal action of the Lie algebra by

$$\mathfrak{g}^c \to \mathrm{Vect}(X) : \zeta \mapsto v_\zeta := v_\xi + Jv_\eta.$$

[1] A subgroup of $\mathrm{GL}(n, \mathbb{C})$ is a Lie subgroup if and only if it is closed as a subset of $\mathrm{GL}(n, \mathbb{C})$.

Here v_ξ is the Hamiltonian vector field of the function H_ξ and $Jv_\eta = \nabla H_\eta$ is the gradient vector field of the function H_η. Since v_ζ is a holomorphic vector field and $\nabla J = 0$, we have

$$[v_\zeta, Jw] = J[v_\zeta, w], \qquad \nabla_{Jw} v_\zeta = J\nabla_w v_\zeta \tag{2.4}$$

for every $\zeta \in \mathfrak{g}^c$ and every vector field $w \in \mathrm{Vect}(X)$.[2]

Alternative Notation for the Infinitesimal Action For each $x \in X$ we use the alternative notation $L_x : \mathfrak{g} \to T_xX$ and $L_x^c : \mathfrak{g}^c \to T_xX$ for the infinitesimal action of the Lie algebras $\mathfrak{g}$ and $\mathfrak{g}^c$. Thus

$$L_x\xi := v_\xi(x), \qquad L_x^c\zeta := v_\zeta(x) = L_x\xi + JL_x\eta \tag{2.5}$$

for $\zeta = \xi + \mathbf{i}\eta \in \mathfrak{g}^c$. Then Eqs. (2.1) and (2.2) take the form

$$L_x{}^* = d\mu(x)J, \qquad d\mu(x)^* = JL_x, \qquad d\mu(x)L_x\xi = -[\mu(x), \xi] \tag{2.6}$$

for $x \in X$ and $\xi \in \mathfrak{g}$.

Lemma 2.1 *Let $x_0, x_1 \in \mu^{-1}(0)$. If $x_1 \in \mathrm{G}^c(x_0)$ then $x_1 \in \mathrm{G}(x_0)$. In fact, if $\eta \in \mathfrak{g}$ and $u \in \mathrm{G}$ satisfy*

$$\exp(\mathbf{i}\eta)ux_0 = x_1$$

then $ux_0 = x_1$ and $L_{x_1}\eta = 0$.

Proof Choose $g_0 \in \mathrm{G}^c$ such that $x_1 = g_0x_0$ and define $\eta \in \mathfrak{g}$ and $u \in \mathrm{G}$ by

$$g_0 =: \exp(\mathbf{i}\eta)u.$$

Define the curve $x : [0, 1] \to X$ by $x(t) := \exp(\mathbf{i}t\eta)ux_0$. Then

$$x(0) = ux_0, \qquad x(1) = x_1, \qquad \dot{x} = JL_x\eta.$$

Hence, by Eq. (2.1),

$$\begin{aligned} \frac{d}{dt}\langle\mu(x), \eta\rangle &= \langle d\mu(x)\dot{x}, \eta\rangle \\ &= \omega(L_x\eta, \dot{x}) \\ &= \omega(L_x\eta, JL_x\eta) \end{aligned} \tag{2.7}$$

[2] We use the sign convention $[v, w] := \nabla_w v - \nabla_v w$ for the Lie bracket of two vector fields so that the infinitesimal action is a homomorphism $\mathfrak{g}^c \to \mathrm{Vect}(X)$ of Lie algebras (and not an anti-homomorphism).

$$= |L_x\eta|^2$$
$$\geq 0.$$

Since $\langle\mu(ux_0),\eta\rangle = \langle\mu(x_1),\eta\rangle = 0$ it follows that $L_{x(t)}\eta = 0$ for all t. Hence $x(t)$ is constant and hence $x_1 = ux_0$. This proves Lemma 2.1. □

Lemma 2.1 asserts that two points in the zero set of the moment map are equivalent under G^c if and only if they are equivalent under G. In notation commonly used in symplectic geometry it says that the Marsden–Weinstein quotient is homeomorphic to the complex quotient, i.e.

$$X/\!\!/G \simeq X^{\mathrm{ps}}/G^c,$$

where $X/\!\!/G := \mu^{-1}(0)/G$ and

$$X^{\mathrm{ps}} := \left\{x \in X \,\middle|\, G^c(x) \cap \mu^{-1}(0) \neq \emptyset\right\}.$$

These spaces can be highly singular. Lemma 2.2 below asserts that $x \in \mu^{-1}(0)$ is a regular point for the moment map if and only if its isotropy subgroup

$$G_x := \{u \in G \,|\, ux = x\}$$

is discrete. Lemma 2.3 below asserts that the complex isotropy subgroup

$$G^c_x := \left\{g \in G^c \,|\, gx = x\right\}$$

is the complexification of G_x whenever $\mu(x) = 0$. This means that there exists an isomorphism of orbifolds

$$X^{\mathrm{s}}/\!\!/G \cong X^{\mathrm{s}}/G^c,$$

where $X^{\mathrm{s}} \subset X^{\mathrm{ps}}$ is the open subset of all points $x \in X^{\mathrm{ps}}$ with discrete isotropy subgroup G^c_x. In general, X^{ps} is not an open subset of X. In geometric invariant theory one studies the quotient $X^{\mathrm{ss}}/\!\!/G^c$, where $X^{\mathrm{ss}} \subset X$ is the open set

$$X^{\mathrm{ss}} := \left\{x \in X \,\middle|\, \overline{G^c(x)} \cap \mu^{-1}(0) \neq \emptyset\right\}$$

and two elements $x, x' \in X^{\mathrm{ss}}$ are equivalent iff $\mu^{-1}(0) \cap \overline{G^c(x)} \cap \overline{G^c(x')} \neq \emptyset$. This quotient is naturally homeomorphic to X^{ps}/G^c and hence to the Marsden–Weinstein quotient $X/\!\!/G$ (see Chap. 7).

Lemma 2.2 *Let $x \in X$ such that $\mu(x) = 0$. Then the following are equivalent.*

(i) *$d\mu(x) : T_x X \to \mathfrak{g}$ is onto.*
(ii) *$L_x : \mathfrak{g} \to T_x X$ is injective.*
(iii) *$L_x^c : \mathfrak{g}^c \to T_x X$ is injective.*

Proof The equation $d\mu(x)J = L_x{}^*$ in (2.6) shows that (i) is equivalent to (ii). That (iii) implies (ii) is obvious. Now assume (ii) and choose an element

$$\zeta = \xi + \mathbf{i}\eta \in \ker L_x^c.$$

Then

$$0 = d\mu(x)\big(L_x\xi + JL_x\eta\big) = -[\mu(x), \xi] + L_x{}^* L_x\eta = L_x{}^* L_x\eta$$

by (2.6). Hence $L_x\eta = 0$, hence

$$L_x\xi = L_x^c\zeta - JL_x\eta = 0,$$

and hence $\xi = \eta = 0$ by (ii). Thus (ii) implies (iii) and this proves Lemma 2.2. □

The hypothesis $\mu(x) = 0$ in Lemma 2.2 cannot be removed. For example a pair of nonantipodal points on the 2-sphere has a trivial isotropy subgroup in SO(3) but a nontrivial isotropy subgroup in the complexified group PSL(2, $\mathbb{C}$). (When the points are antipodal the moment map is zero.)

Lemma 2.3

(i) *Let $x \in X$ such that $\mu(x) = 0$. Then G_x^c is the complexification of G_x, i.e.*

$$\mathrm{G}_x^c = \{\exp(\mathbf{i}\eta)u \mid u \in \mathrm{G}_x,\ \eta \in \ker L_x\}.$$

(ii) *Let $\xi \in \mathfrak{g}$. Then the isotropy subgroup of ξ in G^c is the complexification of the isotropy subgroup in G, i.e. if $\eta \in \mathfrak{g}$, $u \in \mathrm{G}$, and $g := \exp(\mathbf{i}\eta)u$, then*

$$g\xi g^{-1} = \xi \qquad \Longleftrightarrow \qquad u\xi u^{-1} = \xi \quad \text{and} \quad [\xi, \eta] = 0.$$

Proof We prove (i). If $u \in \mathrm{G}$ and $\eta \in \mathfrak{g}$ are such that $g := \exp(\mathbf{i}\eta)u \in \mathrm{G}_x^c$, then Lemma 2.1 asserts that $ux = x$ and $L_x\eta = 0$. This proves (i).

We prove (ii). Assume $g\xi g^{-1} = \xi$, abbreviate

$$h := \exp(-\mathbf{i}\eta),$$

and let

$$\mathbb{C}^n = V_1 \oplus \cdots \oplus V_k$$

be the eigenspace decomposition of ξ. Since

$$h\xi h^{-1} = u\xi u^{-1},$$

it follows that $hV_i \perp hV_j$ for $i \neq j$. Since $h = h^* > 0$ this implies

$$h^2 V_i \subset V_i,$$

hence $hV_i \subset V_i$, and hence

$$uV_i \subset V_i, \qquad \eta V_i \subset V_i$$

for all i. Thus u and η commute with ξ. This proves Lemma 2.3. □

Chapter 3
The Moment Map Squared

Throughout this chapter we denote by $f : X \to \mathbb{R}$ the function defined by

$$f(x) := \tfrac{1}{2}|\mu(x)|^2 \qquad \text{for } x \in X. \tag{3.1}$$

Lemma 3.1 *The gradient of f is given by*

$$\nabla f(x) = JL_x\mu(x)$$

for $x \in X$. Hence $x \in X$ is a critical point of f if and only if $L_x\mu(x) = 0$.

Proof By Eq. (2.1) we have $\langle d\mu(x)\widehat{x}, \xi\rangle = \omega(L_x\xi, \widehat{x})$. Take $\xi = \mu(x)$. Then

$$df(x)\widehat{x} = \langle d\mu(x)\widehat{x}, \mu(x)\rangle = \omega(L_x\mu(x), \widehat{x}) = \langle JL_x\mu(x), \widehat{x}\rangle$$

for $\widehat{x} \in T_xX$. This proves Lemma 3.1. □

By Lemma 3.1 the negative gradient flow line of f through $x_0 \in X$ is the solution of the differential equation.

$$\dot{x} = -JL_x\mu(x), \qquad x(0) = x_0. \tag{3.2}$$

The complexified group orbits are invariant under the gradient flow.

Lemma 3.2 *Let $x_0 \in X$, let $x : \mathbb{R} \to X$ be the unique solution of the differential equation* (3.2)*, and let $g : \mathbb{R} \to \mathrm{G}^c$ be the unique solution of the differential equation*

$$g(t)^{-1}\dot{g}(t) = \mathbf{i}\mu(x(t)), \qquad g(0) = 1. \tag{3.3}$$

© The Author(s), under exclusive license to Springer Nature Switzerland AG 2021
V. Georgoulas et al., *The Moment-Weight Inequality and the Hilbert–Mumford Criterion*, Lecture Notes in Mathematics 2297,
https://doi.org/10.1007/978-3-030-89300-2_3

Then

$$x(t) = g(t)^{-1}x_0$$

for all $t \in \mathbb{R}$.

Proof Define $y : \mathbb{R} \to X$ by $y(t) := g(t)^{-1}x_0$. Since

$$\frac{d}{dt}g^{-1} = -g^{-1}\dot{g}g^{-1}, \qquad g^{-1}\dot{g} = \mathbf{i}\mu(x),$$

it follows that

$$\dot{y} = -g^{-1}\dot{g}g^{-1}x_0 = -L^c_{g^{-1}x_0}(g^{-1}\dot{g}) = -JL_y\mu(x)$$

and $y(0) = x_0$. Hence $x(t) = y(t) = g(t)^{-1}x_0$ for all t. This proves Lemma 3.2. □

Theorem 3.3 (Convergence Theorem) *There exist positive constants* C, c, ε *and*

$$\tfrac{1}{2} < \alpha < 1$$

with the following significance. Let $x_0 \in X$ *and let* $x : \mathbb{R} \to X$ *be the unique solution of* (3.2)*. Then the limit*

$$x_\infty := \lim_{t\to\infty} x(t)$$

exists and satisfies

$$L_{x_\infty}\mu(x_\infty) = 0.$$

Moreover, there exists a constant $T \in \mathbb{R}$ *such that, for all* $t > T$,

$$\begin{aligned} d(x(t), x_\infty) &\leq \int_t^\infty |\dot{x}(s)|\, ds \\ &\leq \frac{C}{1-\alpha}\big(f(x(t)) - f(x_\infty)\big)^{1-\alpha} \\ &\leq \frac{c}{(t-T)^\varepsilon}. \end{aligned} \tag{3.4}$$

The proof of Theorem 3.3 is based on the Lojasiewicz gradient inequality [43, 57], which holds for general analytic gradient flows. That it also applies to the moment map squared was noted by Duistermaat (see Lerman [56] and also Chen–Sun [23, Corollary 3.2]). The result was carried over to certain infinite-dimensional settings by Simon [71] and Morgan–Mrowka–Rubermann [62]. We include a proof for completeness of the exposition.

Proof of Theorem 3.3 Using the Marle–Guillemin–Sternberg local normal form (see [56, Theorem 2.1]), one can show that the moment map is locally real analytic in suitable coordinates. This implies that $f = \frac{1}{2}|\mu|^2$ satisfies the **Lojasiewicz gradient inequality**: There exist constants $\delta > 0$, $C > 0$, and $1/2 < \alpha < 1$ such that, for every critical value a of f and every $x \in X$,

$$|f(x) - a| < \delta \qquad \Longrightarrow \qquad |f(x) - a|^\alpha \leq C|\nabla f(x)|. \tag{3.5}$$

For a proof see Bierstone–Milman [5, Theorem 6.4].[1]

Let $x : \mathbb{R} \to X$ be a nonconstant negative gradient flow line of f. Then

$$a := \lim_{t\to\infty} f(x(t))$$

is a critical value of f. Choose a constant $T \in \mathbb{R}$ such that

$$a < f(x(t)) < a + \delta$$

for $t \geq T$. Then, for $t \geq T$,

$$-\frac{d}{dt}(f(x) - a)^{1-\alpha} = (1-\alpha)(f(x) - a)^{-\alpha}|\nabla f(x)|^2 \geq \frac{1-\alpha}{C}|\dot{x}|.$$

Integrate this inequality over the interval $[t, \infty)$ to obtain

$$\int_t^\infty |\dot{x}(s)|\, ds \leq \frac{C}{1-\alpha}\big(f(x(t)) - a\big)^{1-\alpha} \qquad \text{for } t \geq T. \tag{3.6}$$

This proves the existence of the limit

$$x_\infty := \lim_{t\to\infty} x(t).$$

This limit is a critical point of f and hence satisfies

$$L_{x_\infty}\mu(x_\infty) = 0$$

by Lemma 3.1.

Now abbreviate

$$\rho(t) := (f(x(t)) - a)^{-(2\alpha-1)}.$$

[1] The constants δ, C, α are independent of the choices of a and x because the manifold X is compact by assumption. Care must be taken when extending the result to noncompact manifolds.

Then

$$\dot{\rho}(t) = (2\alpha - 1)\big(f(x(t)) - a\big)^{-2\alpha}|\nabla f(x(t))|^2 \geq \frac{2\alpha - 1}{C^2} \qquad \text{for } t \geq T.$$

This implies

$$\rho(t) \geq \frac{2\alpha - 1}{C^2}(t - T) \qquad \text{for } t \geq T$$

and hence

$$\big(f(x(t)) - a\big)^{1-\alpha} = \rho(t)^{-\frac{1-\alpha}{2\alpha-1}} \leq \left(\frac{2\alpha - 1}{C^2}(t - T)\right)^{-\frac{1-\alpha}{2\alpha-1}}$$

for $t > T$. Thus

$$\frac{C}{1-\alpha}\big(f(x(t)) - a\big)^{1-\alpha} \leq \frac{c}{(t-T)^{\varepsilon}}, \quad \varepsilon := \frac{1-\alpha}{2\alpha - 1}, \quad c := \frac{C}{1-\alpha}\left(\frac{C^2}{2\alpha - 1}\right)^{\varepsilon},$$

for $t > T$. By (3.6), this proves Theorem 3.3. □

We close this section with three lemmas about the Hessian of the moment map squared, culminating in Lijing Wang's inequality in Lemma 3.6. They do not play any role elsewhere in this book.

Lemma 3.4 *Let $x \in X$, $\widehat{x} \in T_x X$, and $\xi, \eta \in \mathfrak{g}$. If $L_x\xi = 0$ then*

$$d\mu(x)\nabla_{\widehat{x}} v_\xi(x) = -[d\mu(x)\widehat{x}, \xi] \tag{3.7}$$

and

$$L_x{}^* L_x[\xi, \eta] = [\xi, L_x{}^* L_x \eta]. \tag{3.8}$$

Proof Differentiate the function

$$\langle [\mu, \xi], \eta\rangle = \langle \mu, [\xi, \eta]\rangle = \omega(v_\xi, v_\eta) = \langle J v_\xi, v_\eta\rangle$$

at x in the direction $\widehat{x}$ and use the equations $v_\xi(x) = 0$ and $\nabla J = 0$ to obtain

$$\begin{aligned}
\langle [d\mu(x)\widehat{x}, \xi], \eta\rangle &= \langle J\nabla_{\widehat{x}} v_\xi(x), v_\eta(x)\rangle \\
&= -\omega(v_\eta(x), \nabla_{\widehat{x}} v_\xi(x)) \\
&= -dH_\eta(x)\nabla_{\widehat{x}} v_\xi(x) \\
&= -\langle d\mu(x)\nabla_{\widehat{x}} v_\xi(x), \eta\rangle.
\end{aligned}$$

This proves (3.7).

It follows from Eqs. (2.4), (2.6), and (3.7) that

$$\begin{aligned}[\xi, L_x{}^*L_x\eta] &= -[d\mu(x)JL_x\eta, \xi] \\ &= d\mu(x)\nabla_{Jv_\eta}v_\xi(x) \\ &= d\mu(x)[v_\xi, Jv_\eta](x) \\ &= d\mu(x)J[v_\xi, v_\eta](x) \\ &= d\mu(x)Jv_{[\xi,\eta]}(x) \\ &= L_x{}^*L_x[\xi, \eta].\end{aligned}$$

This proves (3.8) and Lemma 3.4. □

Lemma 3.5 *The covariant Hessian of*

$$f = \tfrac{1}{2}|\mu|^2$$

at a point $x \in X$ *is the quadratic form* $d^2 f_x : T_xX \to \mathbb{R}$ *given by*

$$d^2 f_x(\widehat{x}) = |d\mu(x)\widehat{x}|^2 + \langle \nabla_{J\widehat{x}}v_\xi(x), \widehat{x}\rangle, \qquad \xi := \mu(x), \tag{3.9}$$

for $\widehat{x} \in T_xX$. *If* x *is a critical point of* f *then*

$$d^2 f_x(JL_x\eta) = |L_x{}^*L_x\eta|^2 - |[\mu(x), \eta]|^2 \tag{3.10}$$

for every $\eta \in \mathfrak{g}$.

Proof Since $\nabla f(x) = JL_x\mu(x) = Jv_{\mu(x)}(x)$ and $\nabla J = 0$, we have

$$\nabla_{\widehat{x}}\nabla f(x) = JL_x d\mu(x)\widehat{x} + J\nabla_{\widehat{x}}v_\xi(x), \qquad \xi := \mu(x).$$

Hence the covariant Hessian of f at x is given by

$$d^2 f_x(\widehat{x}) := \langle \nabla_{\widehat{x}}\nabla f(x), \widehat{x}\rangle = |d\mu(x)\widehat{x}|^2 + \langle \nabla_{J\widehat{x}}v_\xi(x), \widehat{x}\rangle.$$

Here the last step uses the identity $JL_x = d\mu(x)^*$ in (2.6). Take $\widehat{x} := JL_x\eta$ and assume $L_x\mu(x) = 0$ to obtain

$$\begin{aligned}d^2 f_x(JL_x\eta) &= |d\mu(x)JL_x\eta|^2 - \langle \nabla_{L_x\eta}v_\xi(x), JL_x\eta\rangle \\ &= |L_x{}^*L_x\eta|^2 - \langle d\mu(x)\nabla_{L_x\eta}v_\xi(x), \eta\rangle \\ &= |L_x{}^*L_x\eta|^2 + \langle [d\mu(x)L_x\eta, \xi], \eta\rangle \\ &= |L_x{}^*L_x\eta|^2 + \langle d\mu(x)L_x\eta, [\xi, \eta]\rangle \\ &= |L_x{}^*L_x\eta|^2 - |[\mu(x), \eta]|^2.\end{aligned}$$

Here the third step follows from Eq. (3.7) in Lemma 3.4 with $\widehat{x} = L_x\eta$ and the last step follows from Eq. (2.6). This proves Lemma 3.5. □

Lemma 3.6 (Lijing Wang's Inequality) *Let $x \in X$ be a critical point of the moment map squared. Then, for every $\eta \in \mathfrak{g}$,*

$$|[\mu(x), \eta]| \leq |L_x{}^*L_x\eta|. \tag{3.11}$$

Proof The proof is taken from Lijing Wang's paper [77, Theorem 3.8]. Define the linear maps $A, B : \mathfrak{g} \to \mathfrak{g}$ by

$$A\xi := L_x{}^*L_x\xi, \qquad B\xi := [\mu(x), \xi].$$

Thus A is self-adjoint and B is skew-adjoint. Moreover, A and B commute by Lemma 3.4. Identify $\mathfrak{g}^c$ with $\mathfrak{g} \oplus \mathfrak{g}$ and define $P^\pm : \mathfrak{g}^c \to \mathfrak{g}^c$ by

$$P^+ := \begin{pmatrix} A & B \\ -B & A \end{pmatrix}, \qquad P^- := \begin{pmatrix} A & -B \\ B & A \end{pmatrix}.$$

Then the operators P^+ and P^- are self-adjoint, they commute, and

$$\langle \zeta, P^\pm \zeta \rangle = |L_x\xi \pm JL_x\eta|^2.$$

Hence the operator $Q := P^+P^-$ is self-adjoint and nonnegative. Hence

$$0 \leq \langle \xi, Q\xi \rangle = |L_x{}^*L_x\xi|^2 - |[\mu(x), \xi]|^2$$

for $\xi \in \mathfrak{g} \subset \mathfrak{g}^c$. This proves Lemma 3.6. □

Chapter 4
The Kempf–Ness Function

The present chapter introduces the Kempf–Ness function

$$\Phi_x : \mathrm{G}^c/\mathrm{G} \to \mathbb{R}$$

on the Hadamard space G^c/G, associated to an element $x \in X$. In particular, it shows that every gradient flow line of the moment map squared in $\mathrm{G}^c(x)$ gives rise to a gradient flow line of the Kempf–Ness function. It also shows that the Kempf–Ness function is convex along geodesics and that it is a Morse–Bott function.

Equip G^c with the unique left invariant Riemannian metric which agrees with the inner product

$$\langle \xi_1 + \mathbf{i}\eta_1, \xi_2 + \mathbf{i}\eta_2 \rangle_{\mathfrak{g}^c} = \langle \xi_1, \xi_2 \rangle_{\mathfrak{g}} + \langle \eta_1, \eta_2 \rangle_{\mathfrak{g}}$$

on the tangent space $\mathfrak{g}^c$ to G^c at the identity. This metric is invariant under the right G-action. Let

$$\pi : \mathrm{G}^c \to M, \qquad M := \mathrm{G}^c/\mathrm{G}, \tag{4.1}$$

be the projection onto the right cosets of G. This is a principal G-bundle over a contractible manifold. The (orthogonal) splitting

$$\mathfrak{g}^c = \mathfrak{g} \oplus \mathbf{i}\mathfrak{g}$$

extends to a left invariant principal connection on π. The projection from the horizontal bundle (i.e. the summand corresponding to $\mathbf{i}\mathfrak{g}$) defines a G^c-invariant Riemannian metric of nonpositive curvature on M (Appendix C). The geodesic on M through $p = \pi(g)$ in the direction $v = d\pi(g)g\mathbf{i}\xi \in T_pM$ has the form

$$\gamma(t) = \pi(g \exp(\mathbf{i}t\xi)).$$

© The Author(s), under exclusive license to Springer Nature Switzerland AG 2021
V. Georgoulas et al., *The Moment-Weight Inequality and the Hilbert–Mumford Criterion*, Lecture Notes in Mathematics 2297,
https://doi.org/10.1007/978-3-030-89300-2_4

Thus M is complete and, by Hadamard's theorem, diffeomorphic to $\mathbf{i}\mathfrak{g}$ (see also Eq. (2.3)).

Theorem 4.1 (Conjugacy Theorem) *Fix an element $x \in X$.*

(i) *There exists a unique function $\Phi_x : \mathrm{G}^c \to \mathbb{R}$ such that*

$$d\Phi_x(g)\widehat{g} = -\langle \mu(g^{-1}x), \mathrm{Im}(g^{-1}\widehat{g}) \rangle, \qquad \Phi_x(u) = 0, \tag{4.2}$$

for all $g \in \mathrm{G}^c$, all $\widehat{g} \in T_g\mathrm{G}^c$, and all $u \in \mathrm{G}$.

(ii) *Define a map $\psi_x : \mathrm{G}^c \to \mathrm{G}^c(x) \subset X$ by $\psi_x(g) := g^{-1}x$. Then ψ_x intertwines the gradient vector field $\nabla\Phi_x \in \mathrm{Vect}(\mathrm{G}^c)$ and the gradient vector field $\nabla f \in \mathrm{Vect}(X)$, i.e. for all $g \in \mathrm{G}^c$*

$$d\psi_x(g)\nabla\Phi_x(g) = \nabla f(\psi_x(g)). \tag{4.3}$$

Assertion (ii) of Theorem 4.1 is a reformulation of Lemma 3.2 and shows again that ∇f is tangent to the G^c-orbits. Moreover, when the isotropy subgroup of x is discrete, Eqs. (4.2) and (4.3) are equivalent. So in this case the function Φ_x is uniquely determined by (4.3) and the normalization condition $\Phi_x(u) = 0$ for all $u \in \mathrm{G}$. (If G is connected it suffices to impose the condition $\Phi_x(\mathbb{1}) = 0$.) In the opposite case, when x is a fixed point of the group action, Eq. (4.3) carries no information about Φ_x.

Definition 4.2 The function $\Phi_x : \mathrm{G}^c \to \mathbb{R}$ in Theorem 4.1 is called the **lifted Kempf–Ness function** based at x. It is G-invariant and hence descends to a function

$$\Phi_x : M \to \mathbb{R}$$

denoted by the same symbol and called the **Kempf–Ness function**.

Proof of Theorem 4.1 Define a vector field $v_x \in \mathrm{Vect}(\mathrm{G}^c)$ and a 1-form α_x on G^c by

$$v_x(g) := -g\mathbf{i}\mu(g^{-1}x), \qquad \alpha_x(g)\widehat{g} := -\langle \mu(g^{-1}x), \mathrm{Im}(g^{-1}\widehat{g}) \rangle \tag{4.4}$$

for $g \in \mathrm{G}^c$ and $\widehat{g} \in T_g\mathrm{G}^c$. The vector field v_x is horizontal since $\mu(x) \in \mathfrak{g}$ and is right G-equivariant, i.e. $v_x(gu) = v_x(g)u$ for $g \in \mathrm{G}^c$ and $u \in \mathrm{G}$. We must prove the following.

Step 1 The map ψ_x intertwines the vector fields v_x and ∇f, i.e. for all $g \in \mathrm{G}^c$,

$$d\psi_x(g)v_x(g) = \nabla f(\psi_x(g)). \tag{4.5}$$

Step 2 If $\Phi_x : \mathrm{G}^c \to \mathbb{R}$ satisfies $d\Phi_x = \alpha_x$ then its gradient $\nabla\Phi_x$ is v_x, i.e.

$$\alpha_x(g)\widehat{g} = \langle v_x(g), \widehat{g} \rangle_g \tag{4.6}$$

for all $g \in \mathrm{G}^c$ and all $\widehat{g} \in T_g\mathrm{G}^c$. The inner product on the right is the left-invariant Riemannian metric on G^c.

Step 3 There exists a unique function $\Phi_x : \mathrm{G}^c \to \mathbb{R}$ such that $\Phi_x|_{\mathrm{G}} = 0$ and

$$d\Phi_x = \alpha_x. \tag{4.7}$$

We prove Step 1. If $\widehat{g} \in T_g\mathrm{G}^c$ then

$$d\psi_x(g)\widehat{g} = -g^{-1}\widehat{g}g^{-1}x = -L^c_{g^{-1}x}(g^{-1}\widehat{g}).$$

So taking $\widehat{g} = v_x(g) = -g\mathbf{i}\mu(g^{-1}x)$ we get

$$d\psi_x(g)v_x(g) = L^c_{g^{-1}x}(\mathbf{i}\mu(g^{-1}x)) = JL_{g^{-1}x}\mu(g^{-1}x) = \nabla f(g^{-1}x).$$

Here the last equation follows from Lemma 3.1.

We prove Step 2. Let $g \in \mathrm{G}^c$ and $\widehat{g} \in T_g\mathrm{G}^c$. Then

$$\langle v_x(g), \widehat{g}\rangle_g = -\langle \mathbf{i}\mu(g^{-1}x), g^{-1}\widehat{g}\rangle_{\mathfrak{g}^c} = -\langle \mu(g^{-1}x), \mathrm{Im}(g^{-1}\widehat{g})\rangle_{\mathfrak{g}} = \alpha_x(g)\widehat{g}$$

Here the first and last equations follow from the definitions of v_x and α_x in (4.4). This proves Eq. (4.6) and Step 2.

We prove Step 3. By definition α_x is basic, i.e. it vanishes on the tangent vectors $\widehat{g} = g\xi$ (for $\xi \in \mathfrak{g}$) of the group orbit $\pi(g) = g\mathrm{G}$ and it is invariant under the right action of G on G^c. Hence it descends to a 1-form on M. Since M is connected and simply connected, it suffices to prove that α_x is closed, and for this it suffices to prove that the 1-form $g^*\alpha_x$ is closed for every smooth map $g : \mathbb{R}^2 \to \mathrm{G}^c$.

Let s and t be the standard coordinates on $\mathbb{R}^2$ and let $g : \mathbb{R}^2 \to \mathrm{G}^c$ be a smooth map. Define the maps $z : \mathbb{R}^2 \to X$ and $\zeta_s, \zeta_t : \mathbb{R}^2 \to \mathfrak{g}^c$ by

$$z := g^{-1}x, \qquad \zeta_s := g^{-1}\partial_s g, \qquad \zeta_t := g^{-1}\partial_t g.$$

They satisfy $\partial_s z = -L^c_z\zeta_s$, $\partial_t z = -L^c_z\zeta_t$, and $\partial_t\zeta_s - \partial_s\zeta_t = [\zeta_s, \zeta_t]$. Now define

$$\xi_s := \mathrm{Re}(\zeta_s), \qquad \eta_s := \mathrm{Im}(\zeta_s), \qquad \xi_t := \mathrm{Re}(\zeta_t), \qquad \eta_t := \mathrm{Im}(\zeta_t).$$

Then

$$\begin{aligned} \partial_s z &= -L_z\xi_s - JL_z\eta_s, & \partial_t\xi_s - \partial_s\xi_t &= [\xi_s, \xi_t] - [\eta_s, \eta_t], \\ \partial_t z &= -L_z\xi_t - JL_z\eta_t, & \partial_t\eta_s - \partial_s\eta_t &= [\xi_s, \eta_t] + [\eta_s, \xi_t]. \end{aligned} \tag{4.8}$$

Thus the pullback of α_x under g is the 1-form

$$g^*\alpha_x = -\langle \mu(z), \eta_s\rangle\, ds - \langle \mu(z), \eta_t\rangle\, dt.$$

The 1-form $g^*\alpha_x$ is closed if and only if $\partial_t\langle\mu(z),\eta_s\rangle = \partial_s\langle\mu(z),\eta_t\rangle$. Indeed,

$$\begin{aligned}\partial_t\langle\mu(z),\eta_s\rangle - \partial_s\langle\mu(z),\eta_t\rangle &= \langle\mu(z),\partial_t\eta_s - \partial_s\eta_t\rangle + \langle d\mu(z)\partial_t z,\eta_s\rangle - \langle d\mu(z)\partial_s z,\eta_t\rangle \\ &= \langle\mu(z),[\eta_s,\xi_t]\rangle - \langle d\mu(z)(L_z\xi_t + JL_z\eta_t),\eta_s\rangle \\ &\quad + \langle\mu(z),[\xi_s,\eta_t]\rangle + \langle d\mu(z)(L_z\xi_s + JL_z\eta_s),\eta_t\rangle = 0.\end{aligned}$$

Here the second step follows from (4.8) and the last step follows from (2.6). Thus α_x is closed, as claimed, and this proves Step 3 and Theorem 4.1. □

Theorem 4.3 (Properties of the Kempf–Ness Function)

(i) *The Kempf–Ness function*

$$\Phi_x : M \to \mathbb{R}$$

is Morse–Bott and is convex along geodesics.

(ii) *The critical set of Φ_x is a (possibly empty) closed connected submanifold of M. It is given by*

$$\mathrm{Crit}(\Phi_x) = \left\{\pi(g) \in M \mid \mu(g^{-1}x) = 0\right\}. \tag{4.9}$$

(iii) *If the critical manifold of Φ_x is nonempty, then it consists of the absolute minima of Φ_x and every negative gradient flow line of Φ_x converges exponentially to a critical point.*

(iv) *Even if the critical manifold of Φ_x is empty, every negative gradient flow line*

$$\gamma : \mathbb{R} \to M$$

of Φ_x satisfies

$$\lim_{t\to\infty}\Phi_x(\gamma(t)) = \inf_M \Phi_x. \tag{4.10}$$

(The infimum may be minus infinity.)

(v) *The covariant Hessian of Φ_x at a point $\pi(g) \in M$ is the quadratic form*

$$T_{\pi(g)}M \to \mathbb{R} : d\pi(g)\widehat{g} \mapsto |L_{g^{-1}x}\mathrm{Im}(g^{-1}\widehat{g})|^2.$$

(vi) *Let $g : \mathbb{R} \to \mathrm{G}^c$ be a smooth curve. Then the curve*

$$\gamma := \pi \circ g : \mathbb{R} \to M$$

is a negative gradient flow line of Φ_x if and only if g satisfies the differential equation

$$\mathrm{Im}(g^{-1}\dot{g}) = \mu(g^{-1}x). \tag{4.11}$$

(vii) *The Kempf–Ness functions satisfy*

$$\Phi_{h^{-1}x}(h^{-1}g) = \Phi_x(g) - \Phi_x(h)$$

for $x \in X$ *and* $g, h \in \mathrm{G}^c$.

(viii) *Assume the critical manifold of* Φ_x *is nonempty. Let* $g_i \in \mathrm{G}^c$ *be a sequence such that*

$$\sup_i \Phi_x(\pi(g_i)) < \infty.$$

Then there exists a sequence h_i *in the identity component* $\mathrm{G}^c_{x,0}$ *of* G^c_x *such that* $h_i g_i$ *has a convergent subsequence.*

Proof It follows directly from the definition that

$$\nabla\Phi_x(\pi(g)) = -d\pi(g)g\mathbf{i}\mu(g^{-1}x) \tag{4.12}$$

for every $g \in \mathrm{G}^c$. Hence the negative gradient flow lines of Φ_x lift to solutions of Eq. (4.11) and this proves part (vi). It also follows from (4.12) that the critical set of Φ_x is given by (4.9).

We compute the covariant Hessian of Φ_x. Choose a curve $g : \mathbb{R} \to \mathrm{G}^c$ and consider the composition $\gamma := \pi \circ g : \mathbb{R} \to M$. We compute the covariant derivative of the vector field $\nabla\Phi_x$ along this curve, using the formula for the Levi-Civita connection on M in Appendix C. It is given by

$$\nabla_t\nabla\Phi_x(\pi(g)) = d\pi(g)g\zeta, \quad \zeta := \frac{d}{dt}\big(-\mathbf{i}\mu(g^{-1}x)\big) + [\mathrm{Re}(g^{-1}\dot{g}), -\mathbf{i}\mu(g^{-1}x)].$$

Thus $\zeta = \zeta(t) = \mathbf{i}\eta(t)$, where $\eta(t)$ is given by

$$\begin{aligned}
\eta &= -\frac{d}{dt}\mu(g^{-1}x) - [\mathrm{Re}(g^{-1}\dot{g}), \mu(g^{-1}x)] \\
&= -d\mu(g^{-1}x)\frac{d}{dt}g^{-1}x - [\mathrm{Re}(g^{-1}\dot{g}), \mu(g^{-1}x)] \\
&= d\mu(g^{-1}x)g^{-1}\dot{g}g^{-1}x - [\mathrm{Re}(g^{-1}\dot{g}), \mu(g^{-1}x)] \\
&= d\mu(g^{-1}x)L^c_{g^{-1}x}g^{-1}\dot{g} - [\mathrm{Re}(g^{-1}\dot{g}), \mu(g^{-1}x)] \\
&= d\mu(g^{-1}x)L_{g^{-1}x}\mathrm{Re}(g^{-1}\dot{g}) + d\mu(g^{-1}x)JL_{g^{-1}x}\mathrm{Im}(g^{-1}\dot{g}) \\
&\quad - [\mathrm{Re}(g^{-1}\dot{g}), \mu(g^{-1}x)] \\
&= L_{g^{-1}x}{}^*L_{g^{-1}x}\mathrm{Im}(g^{-1}\dot{g}).
\end{aligned}$$

Here the last equation follows from (2.6). Take the inner product of the vector fields $\nabla_t \nabla \Phi_x(\pi(g))$ and $d\pi(g)\dot{g}$ along the curve $\pi(g)$ in M to obtain

$$d^2_{\pi(g)}\Phi_x(d\pi(g)\dot{g}) = \left|L_{g^{-1}x}\mathrm{Im}(g^{-1}\dot{g})\right|^2.$$

This proves the formula for the Hessian of Φ_x in part (v).

We prove that $\mathrm{Crit}(\Phi_x)$ is a submanifold of M with tangent spaces

$$T_{\pi(g)}\mathrm{Crit}(\Phi_x) = \left\{d\pi(g)g\mathbf{i}\eta \mid \eta \in \ker L_{g^{-1}x}\right\}. \tag{4.13}$$

To see this choose an element $g \in \mathrm{G}^c$ such that

$$\mu(g^{-1}x) = 0.$$

By Hadamard's theorem the map $\mathfrak{g} \to M : \eta \mapsto \pi(g\exp(\mathbf{i}\eta))$ is a diffeomorphism. Moreover, for every $\eta \in \mathfrak{g}$, the following are equivalent.

(a) $\pi(g\exp(\mathbf{i}\eta)) \in \mathrm{Crit}(\Phi_x)$.
(b) $\mu(\exp(-\mathbf{i}\eta)g^{-1}x) = 0$.
(c) $\exp(-\mathbf{i}\eta)g^{-1}x \in \mathrm{G}(g^{-1}x)$.
(d) $L_{g^{-1}x}\eta = 0$.

The equivalence of (a) and (b) follows from the formula (4.12) for the gradient of the Kempf–Ness function. The equivalence of (b) and (c) follows from Lemma 2.1. The equivalence of (c) and (d) follows from the fact that the isotropy subgroup $\mathrm{G}^c_{g^{-1}x}$ is the complexification of $\mathrm{G}_{g^{-1}x}$ by Lemma 2.3. Thus we have proved that the set $\mathrm{Crit}(\Phi_x)$ is the image of the linear subspace $\ker L_{g^{-1}x} \subset \mathfrak{g}$ under the diffeomorphism $\mathfrak{g} \to M : \eta \mapsto \pi(g\exp(\mathbf{i}\eta))$. Hence it is a closed connected submanifold of M with the tangent space (4.13) at $\pi(g)$. This proves part (ii) and, by part (v), that Φ_x is Morse–Bott.

Part (v) also shows that the covariant Hessian of Φ_x is everywhere nonnegative. Hence Φ_x is convex along geodesics. Here is an alternative argument. Let $g_0 \in \mathrm{G}^c$ and $\xi \in \mathfrak{g}$ and define $g : \mathbb{R} \to \mathrm{G}^c$ and $y : \mathbb{R} \to X$ by

$$g(t) := g_0\exp(-\mathbf{i}t\xi), \qquad y(t) := g(t)^{-1}x = \exp(\mathbf{i}t\xi)g_0^{-1}x.$$

Then

$$g^{-1}\dot{g} = -\mathbf{i}\xi, \qquad \dot{y} = JL_y\xi.$$

Moreover, the curve $\gamma := \pi \circ g : \mathbb{R} \to M$ is a geodesic and

$$\frac{d}{dt}(\Phi_x \circ \gamma) = -\langle \mu(g^{-1}x), \mathrm{Im}(g^{-1}\dot{g})\rangle = \langle \mu(y), \xi \rangle. \tag{4.14}$$

Hence, as in Eq. (2.7),

$$\frac{d^2}{dt^2}(\Phi_x \circ \gamma) = \frac{d}{dt}\langle \mu(y), \xi \rangle = \langle d\mu(y) J L_y \xi, \xi \rangle = |L_y \xi|^2 \geq 0. \tag{4.15}$$

This shows again that the Kempf–Ness function is convex along geodesics. Thus we have proved assertions (i), (ii), (v), (vi).

We prove parts (iii) and (iv). Let $\gamma_0, \gamma_1 : \mathbb{R} \to M$ be negative gradient flow lines of the Kempf–Ness function Φ_x. Then there exist solutions $g_0, g_1 : \mathbb{R} \to \mathrm{G}^c$ of the differential equation $g_i^{-1}\dot{g}_i = \mathbf{i}\mu(g_i^{-1}x)$ such that $\gamma_0 = \pi \circ g_0$ and $\gamma_1 = \pi \circ g_1$. Define the curves $\eta : \mathbb{R} \to \mathfrak{g}$ and $u : \mathbb{R} \to \mathrm{G}$ by

$$g_1(t) =: g_0(t) \exp(\mathbf{i}\eta(t)) u(t).$$

Then the curve $\beta_t(s) := \pi\big(g_0(t) \exp(\mathbf{i}s\eta(t))\big)$ for $0 \leq s \leq 1$ is the unique geodesic connecting $\gamma_0(t)$ to $\gamma_1(t)$. Hence

$$\rho(t) := d_M(\gamma_0(t), \gamma_1(t)) = |\eta(t)|.$$

This function is nonincreasing by Lemma A.2.

To prove part (iii) assume that $\mu(g_0(0)^{-1}x) = 0$. Then γ_0 is constant and it follows that $\gamma_1([0, \infty))$ is contained in a compact subset of M. Since Φ_x is a Morse–Bott function, this implies that γ_1 converges exponentially to a critical point of Φ_x. This proves part (iii).

To prove part (iv) we argue by contradiction and assume that

$$a := \lim_{t \to \infty} \Phi_x(\gamma_0(t)) > \inf_M \Phi_x.$$

Then $a > -\infty$ and we can choose γ_1 such that

$$\Phi_x(\gamma_1(0)) < a. \tag{4.16}$$

Since the function $\rho = |\eta| : \mathbb{R} \to \mathbb{R}$ is nonincreasing, there exists a constant $C > 0$ such that $|\eta(t)| \leq C$ for all $t \geq 0$. This implies

$$\begin{aligned}
\left.\frac{d}{ds}\right|_{s=0} \Phi_x(\beta_t(s)) &= d\Phi_x(\gamma_0(t))\dot{\beta}_t(0) \\
&= -\langle \mu(g_0(t)^{-1}x), \eta(t) \rangle \\
&\geq -|\mu(g_0(t)^{-1}x)||\eta(t)| \\
&\geq -C|\mu(g_0(t)^{-1}x)|.
\end{aligned}$$

Since the function $\Phi_x \circ \beta_t : [0, 1] \to \mathbb{R}$ is convex it follows that

$$\begin{aligned}\Phi_x(\gamma_1(t)) &= \Phi_x(\beta_t(1)) \\ &\geq \Phi_x(\beta_t(0)) - C|\mu(g_0(t)^{-1}x)| \\ &= \Phi_x(\gamma_0(t)) - C|\mu(g_0(t)^{-1}x)|.\end{aligned}$$

Since $\Phi_x \circ \gamma_0$ is bounded below and

$$\frac{d}{dt}(\Phi_x \circ \gamma_0)(t) = -|\mu(g_0(t)^{-1}x|^2,$$

there exists a sequence $t_i \to \infty$ such that $\lim_{i\to\infty} |\mu(g_0(t_i)^{-1}x)|^2 = 0$. It follows that

$$\lim_{i\to\infty} \Phi_x(\gamma_1(t_i)) \geq \lim_{i\to\infty} \Phi_x(\gamma(t_i)) = a.$$

This contradicts the assumption (4.16). Thus we have proved part (iv).

We prove part (vii). The 1-forms α_x and $\alpha_{h^{-1}x}$ on G^c in (4.4) satisfy

$$\alpha_x(g)\widehat{g} = -\langle \mu(g^{-1}x), \mathrm{Im}(g^{-1}\widehat{g})\rangle = \alpha_{h^{-1}x}(h^{-1}g)h^{-1}\widehat{g}$$

for all $g \in \mathrm{G}^c$ and $\widehat{g} \in T_g\mathrm{G}^c$. Thus the pullback of the 1-form $\alpha_{h^{-1}x} \in \Omega^1(\mathrm{G}^c)$ under the diffeomorphism $\mathrm{G}^c \to \mathrm{G}^c : g \mapsto h^{-1}g$ agrees with α_x. Hence the pullback of the function $\Phi_{h^{-1}x} : \mathrm{G}^c \to \mathbb{R}$ under the same diffeomorphism differs from Φ_x by a constant on each connected component of G^c. Hence assertion (vii) follows from the normalization condition $\Phi_{h^{-1}x}(u) = \Phi_{h^{-1}x}(\mathbb{1}) = 0$.

We prove part (viii) in five steps.

Step 1 *For all* $g \in \mathrm{G}^c$ *the set*

$$N_g := \{\pi(hg) \mid h \in \mathrm{G}^c_{x,0}\} \subset M$$

is closed.

Choose a sequence $h_i \in \mathrm{G}^c_{x,0}$ and an element $\widetilde{g} \in \mathrm{G}^c$ such that the sequence $\pi(h_i g)$ converges to $\pi(\widetilde{g})$ in M. Then there exists a sequence $u_i \in \mathrm{G}$ such that $h_i g u_i$ converges to $\widetilde{g}$ in G^c. Pass to a subsequence so that the limit

$$u := \lim_{i\to\infty} u_i$$

exists in G. Then h_i converges to

$$h := \widetilde{g}u^{-1}g^{-1} \in \mathrm{G}^c_{x,0}$$

and hence $\pi(\widetilde{g}) = \pi(hg) \in N_g$. This proves Step 1.

Step 2 *If $\mu(x) = 0$ then Φ_x is constant on N_g for all $g \in \mathrm{G}^c$.*
It follows from (4.2) that

$$\frac{d}{dt}\Phi_x(h\exp(t\zeta)) = -\langle \mu(x), \mathrm{Im}(\zeta)\rangle = 0$$

for $h \in \mathrm{G}^c_{x,0}$ and $\zeta \in \ker L^c_x$. Hence $\Phi_x(h) = 0$ for all $h \in \mathrm{G}^c_{x,0}$ and so, by part (vii),

$$\Phi_x(h^{-1}g) = \Phi_{h^{-1}x}(h^{-1}g) = \Phi_x(g) - \Phi_x(h) = \Phi_x(g)$$

for all $g \in \mathrm{G}^c$ and all $h \in \mathrm{G}^c_{x,0}$. This proves Step 2.

Step 3 *If $\mu(x) = 0$ then there is a constant $\delta > 0$ such that, for every $\eta \in \mathfrak{g}$,*

$$\eta \perp \ker L_x, \ |\eta| \geq 1 \qquad \Longrightarrow \qquad \Phi_x(\exp(\mathbf{i}\eta)) \geq \delta|\eta|. \tag{4.17}$$

If $\eta \in \mathfrak{g}$ satisfies $L_x\eta \neq 0$, then $\Phi_x(\exp(\mathbf{i}\eta)) > 0$. This follows from the fact that the function

$$\phi_\eta(t) := \Phi_x(\exp(\mathbf{i}t\eta))$$

is convex and satisfies

$$\phi_\eta(0) = 0, \qquad \dot{\phi}_\eta(0) = 0, \qquad \ddot{\phi}_\eta(0) = |L_x\eta|^2 > 0.$$

Now define

$$\delta := \min\{\Phi_x(\exp(\mathbf{i}\eta)) \,|\, \eta \in \mathfrak{g}, \ \eta \perp \ker L_x, \ |\eta| = 1\}.$$

Then $\delta > 0$ and $\phi_\eta(1) \geq \delta$ for every element $\eta \in (\ker L_x)^\perp$ of norm one. Hence the inequality (4.17) follows from the convexity of the functions ϕ_η. This proves Step 3.

Step 4 *If $\mu(x) = 0$ then part (viii) holds.*
Choose a sequence $g_i \in \mathrm{G}^c$ such that $c := \sup_i \Phi_x(g_i) < \infty$. Since N_{g_i} is a closed subset of M by Step 1, there exists a sequence $h_i \in \mathrm{G}^c_{x,0}$ such that

$$r_i := d_M(\pi(h_i g_i), \pi(\mathbb{1})) = \inf_{h \in \mathrm{G}^c_{x,0}} d_M(\pi(hg_i), \pi(\mathbb{1})). \tag{4.18}$$

Choose $\eta_i \in \mathfrak{g}$ and $u_i \in \mathrm{G}$ such that

$$h_i g_i = \exp(\mathbf{i}\eta_i)u_i.$$

If $\xi \in \ker L_x$ then $\exp(-\mathbf{i}\xi) \in \mathrm{G}^c_{x,0}$, and hence $\pi(\exp(-\mathbf{i}\xi)\exp(\mathbf{i}\eta_i)) \in N_{g_i}$. Thus it follows from (4.18) that, for $\xi \in \ker L_x$,

$$\begin{aligned} r_i &= d_M(\pi(\exp(\mathbf{i}\eta_i)), \pi(\mathbb{1})) \\ &\le d_M(\pi(\exp(-\mathbf{i}\xi)\exp(\mathbf{i}\eta_i)), \pi(\mathbb{1})) \\ &= d_M(\pi(\exp(\mathbf{i}\eta_i)), \pi(\exp(\mathbf{i}\xi))). \end{aligned}$$

In other words, for every $\xi \in \ker L_x$, the geodesic

$$\gamma_\xi(t) := \pi(\exp(\mathbf{i}t\xi))$$

in M has minimal distance to the point $\pi(\exp(\mathbf{i}\eta_i)) = \pi(h_i g_i)$ at $t = 0$, and this implies $\langle \eta_i, \xi \rangle = 0$. Thus $\eta_i \perp \ker L_x$ and, if $|\eta_i| \ge 1$, it follows from Step 2 and Step 3 that

$$c \ge \Phi_x(g_i) = \Phi_x(h_i g_i) = \Phi_x(\exp(\mathbf{i}\eta_i)) \ge \delta|\eta_i|.$$

Hence the sequence η_i is bounded, and so the sequence $h_i g_i = \exp(\mathbf{i}\eta_i)u_i$ has a convergent subsequence. This proves Step 4.

Step 5 *We prove part (viii).*

Let $x \in X$ such that the critical manifold of Φ_x is nonempty. Then there is a $g \in \mathrm{G}^c$ such that $\mu(g^{-1}x) = 0$. Choose a sequence $g_i \in \mathrm{G}^c$ such that the sequence $\Phi_x(g_i)$ is bounded. Then by part (vii) so is the sequence

$$\Phi_{g^{-1}x}(g^{-1}g_i) = \Phi_x(g_i) - \Phi_x(g).$$

Hence, by Step 4, there exists a sequence

$$h_i \in \mathrm{G}^c_{g^{-1}x,0}$$

such that $h_i g^{-1} g_i$ has a convergent subsequence. Thus

$$\widetilde{h}_i := g h_i g^{-1} \in \mathrm{G}^c_{x,0}$$

for all i and the sequence $\widetilde{h}_i g_i = g h_i g^{-1} g_i$ has a convergent subsequence. This proves Step 5 and Theorem 4.3.

□

Chapter 5
μ-Weights

The purpose of the present chapter is to introduce Mumford's numerical invariants $w_\mu(x, \zeta)$ associated to an element $x \in X$ and a toral generator $\zeta \in \mathscr{T}^c$. The main result is Mumford's Theorem 5.3 which establishes the invariance of the μ-weights under the diagonal action of the complexified Lie group G^c on $X \times \mathscr{T}^c$ and under Mumford's equivalence relation on $\mathscr{T}^c$. Toral generators and Mumford's equivalence relation are explained in Appendix D.

Introduce the notations

$$\mathscr{T}^c := \left\{ g\xi g^{-1} \mid g \in G^c, \xi \in \mathfrak{g} \setminus \{0\} \right\}$$

and

$$\Lambda := \{\xi \in \mathfrak{g} \setminus \{0\} \mid \exp(\xi) = 1\}, \quad \Lambda^c := \left\{\zeta \in \mathfrak{g}^c \setminus \{0\} \mid \exp(\zeta) = 1\right\}.$$

Definition 5.1 The μ-**weight of a pair**

$$(x, \zeta) \in X \times \mathscr{T}^c$$

is the real number

$$w_\mu(x, \zeta) := \lim_{t \to \infty} \langle \mu(\exp(\mathbf{i}t\zeta)x), \mathrm{Re}(\zeta) \rangle. \tag{5.1}$$

For $\zeta = \xi \in \mathfrak{g} \setminus \{0\}$ the existence of the limit follows from the fact that the function $t \mapsto \langle \mu(\exp(\mathbf{i}t\xi)x), \xi \rangle$ is nondecreasing by (2.7). For general elements $\zeta \in \mathscr{T}^c$ the existence of the limit follows from Lemma 5.4 below.

For $\zeta \in \Lambda^c$ the geometric significance of the μ-weight in terms of a lift of the G^c action to a line bundle over X, under a suitable rationality hypothesis, is explained in Theorem 9.7 below. For $\xi \in \mathfrak{g} \setminus \{0\}$ the next lemma shows that the

© The Author(s), under exclusive license to Springer Nature Switzerland AG 2021
V. Georgoulas et al., *The Moment-Weight Inequality and the Hilbert–Mumford Criterion*, Lecture Notes in Mathematics 2297,
https://doi.org/10.1007/978-3-030-89300-2_5

μ-weight $w_\mu(x,\xi)$ is the asymptotic slope of the Kempf–Ness function Φ_x along the geodesic ray $t \mapsto [\exp(-\mathbf{i}t\xi)]$ as t tends to ∞.

Lemma 5.2 *Let $x \in X$ and $\xi \in \mathfrak{g} \setminus \{0\}$. Then the function $t \mapsto t^{-1}\Phi_x(\exp(-\mathbf{i}t\xi))$ is nondecreasing and*

$$w_\mu(x,\xi) = \lim_{t\to\infty} \frac{\Phi_x(\exp(-\mathbf{i}t\xi))}{t}. \tag{5.2}$$

Proof By Eq. (4.14) we have

$$\Phi_x(\exp(-\mathbf{i}t\xi)) = \int_0^t \langle \mu(\exp(\mathbf{i}s\xi)x), \xi\rangle\, ds \qquad \text{for all } t > 0.$$

Hence it follows from the definition of the weight in (5.1) that

$$w_\mu(x,\xi) = \lim_{t\to\infty} \frac{1}{t}\int_0^t \langle \mu(\exp(\mathbf{i}s\xi)x), \xi\rangle\, ds = \lim_{t\to\infty} \frac{\Phi_x(\exp(-\mathbf{i}t\xi))}{t}.$$

That the function $[0,\infty) \to \mathbb{R} : t \mapsto t^{-1}\Phi_x(\exp(-\mathbf{i}t\xi))$ is nondecreasing follows from the fact that the function $t \to \Phi_x(\exp(-\mathbf{i}t\xi))$ is convex and vanishes at $t = 0$. This proves Lemma 5.2. □

Theorem 5.3 (Mumford)

(i) *The function*

$$w_\mu : X \times \mathscr{T}^c \to \mathbb{R}$$

is G^c*-invariant, i.e.*

$$w_\mu(gx, g\zeta g^{-1}) = w_\mu(x,\zeta)$$

for all $x \in X$, $\zeta \in \mathscr{T}^c$, and $g \in \mathrm{G}^c$.

(ii) *For every $x \in X$ the function*

$$\mathscr{T}^c \to \mathbb{R} : \zeta \mapsto w_\mu(x,\zeta)$$

is constant on the equivalence classes in Theorem D.4, i.e.

$$w_\mu(x, p\zeta p^{-1}) = w_\mu(x,\zeta)$$

for all $\zeta \in \mathscr{T}^c$ and $p \in \mathrm{P}(\zeta)$.

Proof See page 34. □

The proof of Theorem 5.3 is based on Lemmas 5.4 and 5.8 below. The next lemma establishes the existence of the limit in (5.1).

Lemma 5.4 *Let $x_0 \in X$ and $\zeta \in \mathscr{T}^c$. Then the limits*

$$x^{\pm} := \lim_{t\to\pm\infty} \exp(\mathbf{i}t\zeta)x_0 \tag{5.3}$$

exist, the convergence is exponential in t, and $L^c_{x^\pm}\zeta = 0$.

Proof Assume first that $\zeta = \xi \in \mathfrak{g} \setminus \{0\}$ and define the function $x : \mathbb{R} \to X$ by $x(t) := \exp(\mathbf{i}t\xi)x_0$. Then $\dot{x} = JL_x\xi = \nabla H_\xi(x)$ and so x is a gradient flow line of the Hamiltonian function $H_\xi = \langle \mu, \xi \rangle$. Since H_ξ is a Morse–Bott function the limits $x^\pm := \lim_{t\to\pm\infty} \exp(\mathbf{i}t\xi)x_0$ exist, the convergence is exponential in t, and the limit points satisfy $L_{x^\pm}\xi = 0$. This proves the lemma for $\zeta = \xi \in \mathfrak{g} \setminus \{0\}$.

Now let $\zeta \in \mathscr{T}^c$. By Lemma C.4, there is a $g \in \mathrm{G}^c$ such that $g\zeta g^{-1} \in \mathfrak{g}$. By what we have just proved the limits $\widetilde{x}^\pm := \lim_{t\to\pm\infty} \exp(\mathbf{i}tg\zeta g^{-1})gx_0$ exist, the convergence is exponential in t, and $L^c_{\widetilde{x}^\pm}(g\zeta g^{-1}) = 0$. Hence the curve

$$\exp(\mathbf{i}t\zeta)x_0 = g^{-1}\exp(\mathbf{i}tg\zeta g^{-1})gx_0$$

converges to $x^\pm := g^{-1}\widetilde{x}^\pm$ as t tends to $\pm\infty$ and $L^c_{x^\pm}\zeta = g^{-1}L^c_{\widetilde{x}^\pm}(g\zeta g^{-1}) = 0$. This proves Lemma 5.4. □

Remark 5.5 Let $x_0 \in X$ and $\xi \in \mathfrak{g} \setminus \{0\}$. Define $x(t) := \exp(\mathbf{i}t\xi)x_0$ as in the proof of Lemma 5.4 and let $x^\pm := \lim_{t\to\pm\infty} \exp(\mathbf{i}t\xi)x_0$. Then, as in (2.7),

$$\frac{d}{dt}\langle \mu(x), \xi \rangle = |L_x\xi|^2 = |\dot{x}|^2.$$

Integrate this equation to obtain the energy identity

$$E(x) := \int_{-\infty}^{\infty} |\dot{x}(t)|^2\,dt = \langle \mu(x^+), \xi \rangle - \langle \mu(x^-), \xi \rangle = w_\mu(x_0, \xi) + w_\mu(x_0, -\xi).$$

In particular, $w_\mu(x_0, \xi) + w_\mu(x_0, -\xi) \geq 0$.

Remark 5.6 Here is another proof of Lemma 5.4 for $\zeta \in \Lambda^c$. Let $x_0 \in X$ and define the map $z : \mathbb{S} := \mathbb{R}/\mathbb{Z} \times \mathbb{R} \to X$ by

$$z(s, t) := \exp((s + \mathbf{i}t)\zeta)x_0.$$

We prove that z is a finite energy holomorphic curve and

$$E(z) := \int_{\mathbb{S}} |\partial_s z|^2 = w_\mu(x_0, \zeta) + w_\mu(x_0, -\zeta). \tag{5.4}$$

To see this, let $\xi := \mathrm{Re}(\zeta)$ and $\eta := \mathrm{Im}(\zeta)$. Then

$$\partial_s z = L_z\xi + JL_z\eta, \qquad \partial_t z = JL_z\xi - L_z\eta, \tag{5.5}$$

so $\partial_s z + J\partial_t z = 0$ and z is holomorphic. It follows also from (5.5) that

$$\partial_t \langle \mu(z), \xi \rangle = \langle d\mu(z)(JL_z\xi - L_z\eta), \xi \rangle = |L_z\xi|^2 - \langle \mu(z), [\xi, \eta] \rangle,$$
$$\partial_s \langle \mu(z), \eta \rangle = \langle d\mu(z)(L_z\xi + JL_z\eta), \eta \rangle = |L_z\eta|^2 - \langle \mu(z), [\xi, \eta] \rangle.$$

Since $\langle \mu(z), [\xi, \eta] \rangle = \omega(L_z\xi, L_z\eta) = -\langle L_z\xi, JL_z\eta \rangle$, this implies

$$\partial_t \langle \mu(z), \xi \rangle + \partial_s \langle \mu(z), \eta \rangle = |L_z\xi + JL_z\eta|^2 = |\partial_s z|^2.$$

Integrate this identity over $0 \le s \le 1$ to obtain

$$\frac{d}{dt}\int_0^1 \langle \mu(z), \xi \rangle \, ds = \int_0^1 |L_z\xi + JL_z\eta|^2 \, ds = \int_0^1 |\partial_s z|^2 \, ds.$$

Integrate this identity over $-\infty < t < \infty$ to obtain

$$E(z) = \int_{\mathbb{S}} |\partial_s z|^2 = \lim_{t\to\infty}\int_0^1 \langle \mu(z), \xi \rangle \, ds - \lim_{t\to-\infty}\int_0^1 \langle \mu(z), \xi \rangle \, ds. \tag{5.6}$$

The limits on the right exist because the function $t \mapsto \int_0^1 \langle \mu(z(s,t)), \xi \rangle \, ds$ is nondecreasing and bounded. Hence z has finite energy. Hence it follows from the removable singularity theorem for holomorphic curves in [61, Theorem 4.1.2] that the limits $x^\pm := \lim_{t\to\pm\infty} \exp((s + \mathbf{i}t)\zeta)x_0$ exist and the convergence is uniform in s and exponential in t. Since the limit is independent of s it follows again that $\exp(s\zeta)x^\pm = x^\pm$ for all $s \in \mathbb{R}$ and so $L^c_{x^\pm}\zeta = 0$. Moreover,

$$\lim_{t\to\pm\infty}\int_0^1 \langle \mu(z), \xi \rangle \, ds = \langle \mu(x^\pm), \xi \rangle = \pm w_\mu(x_0, \pm\zeta)$$

and hence Eq. (5.4) follows from (5.6).

Lemma 5.7

(i) *For all* $\zeta = \xi + \mathbf{i}\eta \in \mathfrak{g}^c$ *and all* $g \in \mathrm{G}^c$

$$\begin{aligned} |\mathrm{Re}(g\zeta g^{-1})|^2 - |\mathrm{Im}(g\zeta g^{-1})|^2 &= |\xi|^2 - |\eta|^2, \\ \langle \mathrm{Re}(g\zeta g^{-1}), \mathrm{Im}(g\zeta g^{-1}) \rangle &= \langle \xi, \eta \rangle. \end{aligned} \tag{5.7}$$

(ii) *If* $\zeta = \xi + \mathbf{i}\eta \in \mathscr{T}^c$ *then* $|\xi| > |\eta|$ *and* $\langle \xi, \eta \rangle = 0$.

Proof For $g = u \in \mathrm{G}$ Eq. (5.7) follows from the invariance of the inner product on $\mathfrak{g}$. Hence it suffices to assume $g = \exp(\mathbf{i}\widehat{\eta})$ for some element $\widehat{\eta} \in \mathfrak{g}$. Let $\zeta_0 \in \mathfrak{g}^c$ and define the curves $g : \mathbb{R} \to \mathrm{G}^c$ and $\zeta : \mathbb{R} \to \mathfrak{g}^c$ by

$$g(t) := \exp(\mathbf{i}t\widehat{\eta}), \qquad \zeta(t) := g(t)\zeta_0 g(t)^{-1}.$$

Then $\dot{g}g^{-1} = \mathbf{i}\widehat{\eta}$ and $\dot{\zeta} = [\mathbf{i}\widehat{\eta}, \zeta]$. Define $\xi(t) := \mathrm{Re}(\zeta(t))$ and $\eta(t) := \mathrm{Im}(\zeta(t))$ so that

$$\dot{\xi} = -[\widehat{\eta}, \eta], \qquad \dot{\eta} = [\widehat{\eta}, \xi].$$

Then

$$\frac{d}{dt}\frac{|\xi|^2}{2} = \langle \dot{\xi}, \xi \rangle = -\langle [\widehat{\eta}, \eta], \xi \rangle = \langle [\widehat{\eta}, \xi], \eta \rangle = \langle \dot{\eta}, \eta \rangle = \frac{d}{dt}\frac{|\eta|^2}{2}$$

and

$$\frac{d}{dt}\langle \xi, \eta \rangle = \langle \dot{\xi}, \eta \rangle + \langle \xi, \dot{\eta} \rangle = -\langle [\widehat{\eta}, \eta], \eta \rangle + \langle \xi, [\widehat{\eta}, \xi] \rangle = 0.$$

Hence the functions $t \mapsto |\xi(t)|^2 - |\eta(t)|^2$ and $t \mapsto \langle \xi(t), \eta(t) \rangle$ are constant. This proves part (i).

Now let $\zeta = \xi + \mathbf{i}\eta \in \mathscr{T}^c$. By Lemma C.4 there exists an element $g \in \mathrm{G}^c$ such that $g\zeta g^{-1} \in \mathfrak{g} \setminus \{0\}$. Hence $\mathrm{Im}(g\zeta g^{-1}) = 0$ and it follows from part (i) that

$$|\xi|^2 - |\eta|^2 = |g\zeta g^{-1}|^2 > 0, \qquad \langle \xi, \eta \rangle = 0.$$

This proves part (ii) and Lemma 5.7. □

Lemma 5.8

(i) *If $x \in X$ and $\zeta \in \mathfrak{g}^c$ satisfy $L_x^c \zeta = 0$ then, for all $g \in \mathrm{G}^c$,*

$$\begin{aligned} \langle \mu(gx), \mathrm{Re}(g\zeta g^{-1}) \rangle &= \langle \mu(x), \mathrm{Re}(\zeta) \rangle, \\ \langle \mu(gx), \mathrm{Im}(g\zeta g^{-1}) \rangle &= \langle \mu(x), \mathrm{Im}(\zeta) \rangle. \end{aligned} \tag{5.8}$$

(ii) *If $x \in X$ and $\zeta \in \mathscr{T}^c$ satisfy $L_x^c \zeta = 0$ then $\langle \mu(x), \mathrm{Im}(\zeta) \rangle = 0$.*

Proof For $g = u \in \mathrm{G}$ Eq. (5.8) follows from the invariance of the inner product on $\mathfrak{g}$ and the G-equivariance of the moment map. Hence it suffices to assume that $g = \exp(\mathbf{i}\widehat{\eta})$ for sme $\widehat{\eta} \in \mathfrak{g}$. Fix two elements $x_0 \in X$ and $\zeta_0 \in \mathfrak{g}^c$ such that $L_{x_0}^c \zeta_0 = 0$, define the curves $g : \mathbb{R} \to \mathrm{G}^c$, $x : \mathbb{R} \to X$, and $\zeta : \mathbb{R} \to \mathfrak{g}^c$ by

$$g(t) := \exp(\mathbf{i}t\widehat{\eta}), \qquad x(t) := g(t)x_0, \qquad \zeta(t) := g(t)\zeta_0 g(t)^{-1},$$

and denote $\xi(t) := \mathrm{Re}(\zeta(t))$ and $\eta(t) := \mathrm{Im}(\zeta(t))$. Then $L_x\xi + JL_x\eta = 0$ and, as in the proof of Lemma 5.7,

$$\dot{\xi} = -[\widehat{\eta}, \eta], \qquad \dot{\eta} = [\widehat{\eta}, \xi], \qquad \dot{x} = JL_x\widehat{\eta}.$$

Hence, by Eqs. (2.1) and (2.2),

$$\begin{aligned}\frac{d}{dt}\langle\mu(x),\xi\rangle &= \langle d\mu(x)\dot{x},\xi\rangle + \langle\mu(x),\dot{\xi}\rangle \\ &= \omega(L_x\xi,\dot{x}) + \langle\mu(x),\dot{\xi}\rangle \\ &= -\omega(JL_x\eta,\dot{x}) + \langle\mu(x),\dot{\xi}\rangle \\ &= -\omega(JL_x\eta, JL_x\widehat{\eta}) - \langle\mu(x),[\widehat{\eta},\eta]\rangle \\ &= \omega(L_x\widehat{\eta}, L_x\eta) - \langle\mu(x),[\widehat{\eta},\eta]\rangle \\ &= 0\end{aligned}$$

and

$$\begin{aligned}\frac{d}{dt}\langle\mu(x),\eta\rangle &= \langle d\mu(x)\dot{x},\eta\rangle + \langle\mu(x),\dot{\eta}\rangle \\ &= \omega(L_x\eta,\dot{x}) + \langle\mu(x),\dot{\eta}\rangle \\ &= \omega(JL_x\xi,\dot{x}) + \langle\mu(x),\dot{\eta}\rangle \\ &= \omega(JL_x\xi, JL_x\widehat{\eta}) + \langle\mu(x),[\widehat{\eta},\xi]\rangle \\ &= \omega(L_x\xi, L_x\widehat{\eta}) - \langle\mu(x),[\xi,\widehat{\eta}]\rangle \\ &= 0.\end{aligned}$$

Hence the functions $t \mapsto \langle\mu(x(t)),\xi(t)\rangle$ and $t \mapsto \langle\mu(x(t)),\eta(t)\rangle$ are constant. This proves (i). To prove (ii), let $x \in X$ and $\zeta \in \mathscr{T}^c$ such that $L_x^c\zeta = 0$. Then it follows from Lemma C.4 that there exists an element $g \in \mathrm{G}^c$ such that $g\zeta g^{-1} \in \mathfrak{g}$. By (i) this implies $\langle\mu(x),\mathrm{Im}(\zeta)\rangle = \langle\mu(gx),\mathrm{Im}(g\zeta g^{-1})\rangle = 0$. This proves Lemma 5.8. □

Proof of Theorem 5.3 Let $x \in X$, $\zeta \in \mathscr{T}^c$, and $g \in \mathrm{G}^c$. By Lemma 5.4, the limit

$$x^+ := \lim_{t\to\infty} \exp(\mathbf{i}t\zeta)x$$

exists and satisfies $L_{x^+}^c\zeta = 0$. Moreover,

$$gx^+ = \lim_{t\to\infty} \exp(\mathbf{i}tg\zeta g^{-1})gx.$$

Hence, by Definition 5.1 and Lemma 5.8,

$$w_\mu(x,\zeta) = \langle\mu(x^+),\mathrm{Re}(\zeta)\rangle = \langle\mu(gx^+),\mathrm{Re}(g\zeta g^{-1})\rangle = w_\mu(gx, g\zeta g^{-1}).$$

This proves part (i) of Theorem 5.3.

Now assume, in addition, that $g \in \mathrm{P}(\zeta)$. Then the limit

$$g^+ := \lim_{t\to\infty} \exp(\mathbf{i}t\zeta)g\exp(-\mathbf{i}t\zeta)$$

exists in G^c. It satisfies $\exp(\mathbf{i}s\zeta)g^+\exp(-\mathbf{i}s\zeta) = g^+$ for all $s \in \mathbb{R}$. Differentiate this equation to obtain $\zeta g^+ - g^+\zeta = 0$ and hence

$$\zeta = g^+\zeta(g^+)^{-1}.$$

Moreover,

$$\begin{aligned} g^+x^+ &= \lim_{t\to\infty} \exp(\mathbf{i}t\zeta)g\exp(-\mathbf{i}t\zeta) \cdot \lim_{t\to\infty} \exp(\mathbf{i}t\zeta)x \\ &= \lim_{t\to\infty} \exp(\mathbf{i}t\zeta)gx. \end{aligned}$$

Hence it follows from Definition 5.1 and part (i) of Theorem 5.3 (already proved) that

$$\begin{aligned} w_\mu(x, g^{-1}\zeta g) &= w_\mu(gx, \zeta) \\ &= \langle \mu(g^+x^+), \mathrm{Re}(\zeta)\rangle \\ &= \langle \mu(g^+x^+), \mathrm{Re}(g^+\zeta(g^+)^{-1})\rangle \\ &= \langle \mu(x^+), \mathrm{Re}(\zeta)\rangle \\ &= w_\mu(x, \zeta). \end{aligned}$$

Here the fourth equality follows from Lemma 5.8 and the fact that $L^c_{x^+}\zeta = 0$. Since $P(\zeta)$ is a group (take $p = g^{-1}$) this proves part (ii) of Theorem 5.3. □

Chapter 6
The Moment-Weight Inequality

This chapter is devoted to the proof of the moment-weight inequality which relates the Mumford numerical invariants $w_\mu(x, \xi)$ to the norm of the moment map on the complexified group orbit of x. We begin by proving the moment-weight inequality in a special case and hope that the proof might be of some interest in its own right. The general moment-weight inequality is proved in Theorem 6.7 below.

Theorem 6.1 (Restricted Moment-Weight Inequality) *Let $x \in X$ and fix a toral generator $\zeta = \xi + \mathbf{i}\eta \in \mathscr{T}^c$ such that*

$$L_x\xi + JL_x\eta = 0. \tag{6.1}$$

Then $\langle \mu(x), \eta \rangle = 0$, $|\xi| > |\eta|$, and

$$\frac{\langle \mu(x), \xi \rangle^2}{|\xi|^2 - |\eta|^2} \leq |\mu(gx)|^2 \qquad \text{for all } g \in \mathrm{G}^c. \tag{6.2}$$

Proof The equation $\langle \mu(x), \eta \rangle = 0$ was proved in Lemma 5.8 and the inequality $|\xi| > |\eta|$ was proved in Lemma 5.7. This shows that the quotient on the left in (6.2) is well defined. The estimate (6.2) holds obviously when $\mu(x) = 0$. So assume $\mu(x) \neq 0$. Under this assumption we prove (6.2) in three steps.

Step 1 *The inequality* (6.2) *holds when $L_x\xi = L_x\eta = 0$ and $g = 1$.*
By assumption and Eq. (2.6),

$$[\mu(x), \xi] = -d\mu(x)L_x\xi = 0, \qquad [\mu(x), \eta] = -d\mu(x)L_x\eta = 0.$$

Thus $\zeta = \xi + \mathbf{i}\eta$ commutes with $\mu(x)$. Hence $\zeta - \lambda\mu(x) \in \mathscr{T}^c$ for all $\lambda \in \mathbb{R}$ (such that $\zeta \neq \lambda\mu(x)$). By part (ii) of Lemma 5.7, this implies

$$|\xi - \lambda\mu(x)|^2 \geq |\eta|^2$$

© The Author(s), under exclusive license to Springer Nature Switzerland AG 2021
V. Georgoulas et al., *The Moment-Weight Inequality and the Hilbert–Mumford Criterion*, Lecture Notes in Mathematics 2297,
https://doi.org/10.1007/978-3-030-89300-2_6

for all $\lambda \in \mathbb{R}$. The expression on the left is minimized for $\lambda = |\mu(x)|^{-2}\langle\mu(x), \xi\rangle$. Hence $|\xi|^2 - |\mu(x)|^{-2}\langle\mu(x), \xi\rangle^2 \geq |\eta|^2$ and this is equivalent to (6.2).

Step 2 *The inequality* (6.2) *holds when* $L_x^c\zeta = 0$ *and* $g = 1$.

Choose elements $x_0 \in X$ and $\zeta_0 = \xi_0 + \mathbf{i}\eta_0 \in \mathscr{T}^c$ such that $L_{x_0}\xi_0 + JL_{x_0}\eta_0 = 0$. Let $x : \mathbb{R} \to X$ be the solution of (3.2) and $g : \mathbb{R} \to \mathrm{G}^c$ be the solution of (3.3) so that

$$g(t)^{-1}\dot{g}(t) = \mathbf{i}\mu(x(t)), \qquad x(t) = g(t)^{-1}x_0$$

for every $t \in \mathbb{R}$. Define the curve $\zeta : \mathbb{R} \to \mathscr{T}^c$ by

$$\zeta(t) := \xi(t) + \mathbf{i}\eta(t) := g(t)^{-1}\zeta_0 g(t).$$

Then

$$L_x\xi + JL_x\eta = 0, \qquad \dot{\zeta} = [\zeta, \mathbf{i}\mu(x)], \qquad \dot{\xi} = -[\eta, \mu(x)], \qquad \dot{\eta} = [\xi, \mu(x)].$$

By Eq. (2.6), this implies

$$0 = d\mu(x)(L_x\xi + JL_x\eta) = -[\mu(x), \xi] + L_x{}^*L_x\eta,$$
$$0 = d\mu(x)(-L_x\eta + JL_x\xi) = [\mu(x), \eta] + L_x{}^*L_x\xi,$$

and hence

$$[\mu(x), \xi] = L_x{}^*L_x\eta, \qquad [\mu(x), \eta] = -L_x{}^*L_x\xi.$$

Thus

$$\frac{d}{dt}\frac{|\xi|^2}{2} = \langle\dot{\xi}, \xi\rangle = -\langle[\eta, \mu(x)], \xi\rangle = -\langle\eta, [\mu(x), \xi]\rangle = -|L_x\eta|^2$$

and

$$\frac{d}{dt}\frac{|\eta|^2}{2} = \langle\dot{\eta}, \eta\rangle = \langle[\xi, \mu(x)], \eta\rangle = \langle\xi, [\mu(x), \eta]\rangle = -|L_x\xi|^2.$$

This shows that the integral $\int_0^\infty(|L_x\xi|^2 + |L_x\eta|^2)\,dt$ is finite. Hence there exists a sequence $t_i \to \infty$ such that $L_{x(t_i)}\xi(t_i)$ and $L_{x(t_i)}\eta(t_i)$ converge to zero. Passing to a subsequence, if necessary, we may assume that the limits

$$\xi_\infty := \lim_{i\to\infty}\xi(t_i), \qquad \eta_\infty := \lim_{i\to\infty}\eta(t_i), \qquad x_\infty := \lim_{i\to\infty}x(t_i)$$

exist. These limits satisfy

$$L_{x_\infty}\xi_\infty = L_{x_\infty}\eta_\infty = 0.$$

Since each set of toral generators with given eigenvalues and multiplicities is a closed subset of $\mathfrak{g}^c$, we also have $\xi_\infty + \mathbf{i}\eta_\infty \in \mathscr{T}^c$. Hence it follows from Lemmas 5.7, 5.8, and Step 1 that

$$\frac{\langle \mu(x_0), \xi_0 \rangle^2}{|\xi_0|^2 - |\eta_0|^2} = \frac{\langle \mu(x_\infty), \xi_\infty \rangle^2}{|\xi_\infty|^2 - |\eta_\infty|^2} \le |\mu(x_\infty)|^2 \le |\mu(x_0)|^2.$$

This proves Step 2.

Step 3 *The inequality* (6.2) *holds when* $L_x^c\zeta = 0$.

Let $x \in X$, $\zeta = \xi + \mathbf{i}\eta \in \mathscr{T}^c$, and $g \in \mathrm{G}^c$ be as in the hypotheses of the theorem. Then, by Lemmas 5.7, 5.8, and Step 2,

$$\frac{\langle \mu(x), \xi \rangle^2}{|\xi|^2 - |\eta|^2} = \frac{\langle \mu(gx), \mathrm{Re}(g\zeta g^{-1}) \rangle^2}{|\mathrm{Re}(g\zeta g^{-1})|^2 - |\mathrm{Im}(g\zeta g^{-1})|^2} \le |\mu(gx)|^2.$$

This proves Step 3 and Theorem 6.1.

□

Corollary 6.2 (Kirwan–Ness Inequality) *Let $x \in X$ be a critical point of the moment map squared. Then*

$$|\mu(x)| \le |\mu(gx)| \tag{6.3}$$

for all $g \in \mathrm{G}^c$.

Proof Assume $\mu(x) \neq 0$. Then

$$\xi := \mu(x) \in \mathscr{T}^c, \qquad L_x\xi = 0.$$

Hence it follows from Theorem 6.1 that

$$|\mu(x)|^2 = \frac{\langle \mu(x), \xi \rangle^2}{|\xi|^2} \le |\mu(gx)|^2$$

for all $g \in \mathrm{G}^c$. This proves Corollary 6.2. □

The inequality (6.3) is implicitly contained in the work of Kirwan [53]. For linear actions on projective space it is proved in Ness [66, Theorem 6.2 (i)]. The Kirwan–Ness inequality implies that the Hessian of

$$f = \tfrac{1}{2}|\mu|^2$$

is nonnegative on the subspace

$$\mathrm{im}\, JL_x \subset T_x X$$

for every critical point of f. This is equivalent to Lijing Wang's inequality in Lemma 3.6.

Theorem 6.3 (First Ness Uniqueness Theorem) *Let $x_0, x_1 \in X$ be critical points of the moment map squared. Then*

$$x_1 \in \mathrm{G}^c(x_0) \qquad \Longrightarrow \qquad x_1 \in \mathrm{G}(x_0).$$

For linear actions on projective space this is Theorem 7.1 in Ness [66]. Her proof carries over to the general case and is reproduced below as proof 1. Proof 2 is due to Calabi–Chen [9, Corollary 4.1] who used this argument to establish a uniqueness result for extremal Kähler metrics. The first proof uses the Kirwan–Ness inequality in Corollary 6.2, the second proof does not.

Proof 1 The proof has three steps. By Lemma 2.1 we may assume that $\mu(x_0)$ and $\mu(x_1)$ are nonzero.

Step 1 *We may assume without loss of generality that there exists an element $q \in \mathrm{G}^c$ such that*

$$x_1 = qx_0, \qquad q\mu(x_0)q^{-1} = \mu(x_0).$$

Choose an element $g \in \mathrm{G}^c$ such that

$$x_1 = gx_0.$$

By Theorem D.3 there exists an element $p \in \mathrm{P}(-\mu(x_0))$ such that

$$u := pg^{-1} \in \mathrm{G}$$

and hence $g = u^{-1}p$. Assume without loss of generality that $u = \mathbb{1}$. (Replace x_1 by ux_1 if necessary.) Thus $x_1 = px_0$ and the limit

$$q := \lim_{t\to\infty} \exp(-\mathbf{i}t\mu(x_0))p\exp(\mathbf{i}t\mu(x_0))$$

exists in G^c and commutes with $\mu(x_0)$. Define the curve $x : \mathbb{R} \to X$ by

$$x(t) := \exp(-\mathbf{i}t\mu(x_0))x_1.$$

Since $L_{x_0}\mu(x_0) = 0$, we have

$$\lim_{t\to\infty} x(t) = \lim_{t\to\infty} \exp(-\mathbf{i}t\mu(x_0))p\exp(\mathbf{i}t\mu(x_0))x_0 = qx_0$$

and hence

$$\begin{aligned} w_\mu(x_1, -\mu(x_0)) &= \lim_{t\to\infty} \langle \mu(x(t)), -\mu(x_0)\rangle \\ &= -\langle \mu(qx_0), \mu(x_0)\rangle \\ &= -\langle \mu(qx_0), \mathrm{Re}(q\mu(x_0)q^{-1})\rangle \\ &= -|\mu(x_0)|^2. \end{aligned}$$

The last equation uses Lemma 5.8. Since $t \mapsto \langle \mu(x(t)), -\mu(x_0)\rangle$ is nondecreasing and $x(0) = x_1$, this implies $-\langle \mu(x_1), \mu(x_0)\rangle \le -|\mu(x_0)|^2$ and thus

$$\langle \mu(x_0), \mu(x_1)\rangle \ge |\mu(x_0)|^2 = |\mu(x_1)|^2.$$

(Here the last equation follows from Corollary 6.2.) Hence $\mu(x_0) = \mu(x_1)$. Since $L_{x_1}\mu(x_1) = 0$ it follows that the curve $x(t)$ is constant and hence

$$x_1 = x(0) = \lim_{t\to\infty} x(t) = qx_0.$$

This proves Step 1.

Step 2 *We may assume without loss of generality that there exists an element* $\eta \in \mathfrak{g}$ *such that* $x_1 = \exp(\mathbf{i}\eta)x_0$ *and* $[\eta, \mu(x_0)] = 0$.

Let $q \in \mathrm{G}^c$ be as in Step 1 and choose elements $u \in \mathrm{G}$ and $\eta \in \mathfrak{g}$ such that

$$q = \exp(\mathbf{i}\eta)u.$$

Since $q\mu(x_0)q^{-1} = \mu(x_0)$ it follows from part (ii) of Lemma 2.3 that

$$u\mu(x_0)u^{-1} = \mu(x_0), \qquad [\eta, \mu(x_0)] = 0.$$

Replacing x_1 by $u^{-1}x_1$ and η by $u^{-1}\eta u$, if necessary, we may assume without loss of generality that $u = \mathbb{1}$. This proves Step 2.

Step 3 *We prove Theorem 6.3.*

Let $\eta \in \mathfrak{g}$ be as in Step 2. Define the curve $x : \mathbb{R} \to X$ by

$$x(t) := \exp(\mathbf{i}t\eta)x_0.$$

Then $\dot{x} = JL_x\eta$ and, as in Eq. (2.7),

$$\frac{d}{dt}\langle \mu(x), \eta\rangle = \langle d\mu(x)\dot{x}, \eta\rangle = |L_x\eta|^2 \ge 0. \tag{6.4}$$

Define the vector field $\widehat{x}(t) \in T_{x(t)}X$ along x by

$$\widehat{x}(t) := v_{\mu(x_0)}(x(t)) = L_{x(t)}\mu(x_0).$$

Since $[\eta, \mu(x_0)] = 0$ the Lie bracket $[v_\eta, v_{\mu(x_0)}]$ vanishes, and so by (2.4)

$$\nabla_t \widehat{x} = \nabla_{\dot{x}} v_{\mu(x_0)}(x) = J\nabla_{v_\eta(x)} v_{\mu(x_0)}(x) = J\nabla_{\widehat{x}} v_\eta(x), \qquad \widehat{x}(0) = 0.$$

Hence $L_{x(t)}\mu(x_0) = 0$ for all t and hence

$$\frac{d}{dt}\langle \mu(x), \mu(x_0)\rangle = \langle d\mu(x)JL_x\eta, \mu(x_0)\rangle = \langle L_x\eta, L_x\mu(x_0)\rangle = 0.$$

Since $x(0) = x_0$ and $x(1) = x_1$ this implies

$$\langle \mu(x_1), \mu(x_0)\rangle = |\mu(x_0)|^2 = |\mu(x_1)|^2. \tag{6.5}$$

Here the last equation follows from Corollary 6.2. It follows from (6.5) that

$$\mu(x_1) = \mu(x_0)$$

and hence $\langle \mu(x_1), \eta\rangle = \langle \mu(x_0), \eta\rangle$. By (6.4) this implies

$$L_{x(t)}\eta = 0$$

for all t, hence $L_{x_0}\eta = 0$, and hence $x_1 = x_0$. This proves Theorem 6.3.

□

Proof 2 Choose $g_0 \in \mathrm{G}^c$ such that

$$x_1 = g_0^{-1}x_0. \tag{6.6}$$

Since x_0 and x_1 are critical points of the moment map squared, they satisfy

$$L_{x_0}\mu(x_0) = 0, \qquad L_{x_1}\mu(x_1) = 0. \tag{6.7}$$

Define the curves $g, \widetilde{g} : \mathbb{R} \to \mathrm{G}^c$ by

$$g(t) := \exp(\mathbf{i}t\mu(x_0)), \qquad \widetilde{g}(t) := g_0 \exp(\mathbf{i}t\mu(x_1)).$$

Thus the curves $\gamma := \pi \circ g : \mathbb{R} \to M$ and $\widetilde{\gamma} := \pi \circ \widetilde{g} : \mathbb{R} \to M$ are both geodesics. Moreover, $g(t)^{-1}x_0 = x_0$ and $\widetilde{g}(t)^{-1}x_0 = x_1$ for all t by (6.6) and (6.7). Thus g and $\widetilde{g}$ satisfy the differential equation $g^{-1}\dot{g} = \mathbf{i}\mu(g^{-1}x_0)$. Hence it follows from Theorem 4.3 that γ and $\widetilde{\gamma}$ are negative gradient flow lines of the Kempf–Ness function Φ_{x_0}. Now define $\eta(t) \in \mathfrak{g}$ and $u(t) \in \mathrm{G}$ by $g(t)\exp(\mathbf{i}\eta(t))u(t) := \widetilde{g}(t)$. Then

$$x_1 = \widetilde{g}(t)^{-1}g(t)x_0 = u(t)^{-1}\exp(-\mathbf{i}\eta(t))x_0 \tag{6.8}$$

and, by Theorem C.1,

$$\rho(t) := d_M(\gamma(t), \widetilde{\gamma}(t)) = |\eta(t)|$$

for all t. By Lemma A.2 this function is nonincreasing. If $\rho \equiv 0$ then $x_1 \in \mathrm{G}(x_0)$ by assumption. Hence assume $\rho \not\equiv 0$ and, for each t, denote by $\gamma(\cdot, t) : [0, 1] \to M$ the unique geodesic from $\gamma(0, t) = \gamma(t)$ to $\gamma(1, t) = \widetilde{\gamma}(t)$. Then

$$\gamma(s, t) = \pi(g(t)\exp(\mathbf{i}s\eta(t))) \qquad \text{for } 0 \leq s \leq 1.$$

Hence Eq. (A.2) in Lemma A.2 asserts that

$$\begin{aligned}\dot{\rho}(t) &= -\frac{1}{\rho(t)}\int_0^1 \frac{\partial^2}{\partial s^2}\Phi_{x_0}(g(t)\exp(\mathbf{i}s\eta(t)))\,ds \\ &= -\frac{1}{\rho(t)}\int_0^1 |L_{\exp(-\mathbf{i}s\eta(t))x_0}\eta(t)|^2\,ds.\end{aligned} \tag{6.9}$$

Choose a sequence $t_i \to \infty$ such that the limits

$$\lim_{i\to\infty}\dot{\rho}(t_i) = 0, \qquad \lim_{i\to\infty}\eta(t_i) =: \eta_\infty, \qquad \lim_{i\to\infty}u(t_i) =: u_\infty$$

exist. Then $L_{x_0}\eta_\infty = 0$ by (6.9). Hence it follows from (6.8) that

$$x_1 = \lim_{i\to\infty}u(t_i)^{-1}\exp(-\mathbf{i}\eta(t_i))x_0 = u_\infty^{-1}\exp(-\mathbf{i}\eta_\infty)x_0 = u_\infty^{-1}x_0.$$

Thus $x_1 \in \mathrm{G}(x_0)$ and this completes the second proof of Theorem 6.3. □

Theorem 6.4 (Moment Limit Theorem) *Let $x_0 \in X$ and let $x : \mathbb{R} \to X$ be the solution of* (3.2). *Define $x_\infty := \lim_{t\to\infty} x(t)$. Then*

$$|\mu(x_\infty)| = \inf_{g\in\mathrm{G}^c}|\mu(gx_0)|.$$

Moreover, the G*-orbit of x_∞ depends only on the* G^c*-orbit of x_0.*

Proof The limit x_∞ exists by Theorem 3.3. Moreover, by Lemma 3.2, the solution $x : \mathbb{R} \to X$ of Eq. (3.2) is given by

$$x(t) = g(t)^{-1}x_0,$$

where $g : \mathbb{R} \to \mathrm{G}^c$ is the solution of (3.3). Fix an element $g_0 \in \mathrm{G}^c$ and let $\widetilde{x} : \mathbb{R} \to X$ and $\widetilde{g} : \mathbb{R} \to \mathrm{G}^c$ be the solutions of the differential equations

$$\dot{\widetilde{x}} = -JL_{\widetilde{x}}\mu(\widetilde{x}), \qquad \widetilde{x}(0) = g_0^{-1}x_0,$$

and

$$\widetilde{g}^{-1}\dot{\widetilde{g}} = \mathbf{i}\mu(\widetilde{x}), \qquad \widetilde{g}(0) = g_0.$$

Define $\eta : \mathbb{R} \to \mathfrak{g}$ and $u : \mathbb{R} \to \mathrm{G}$ by

$$\widetilde{g}(t) =: g(t)\exp(\mathbf{i}\eta(t))u(t).$$

Then, by Lemma 3.2,

$$\widetilde{x}(t) = \widetilde{g}(t)^{-1}x_0 = u(t)^{-1}\exp(-\mathbf{i}\eta(t))x(t)$$

for all $t \in \mathbb{R}$. Denote by $d_M : M \times M \to [0, \infty)$ the distance function of the Riemannian metric on the homogeneous space M. By Theorem 4.3 the curves

$$\gamma = \pi \circ g : \mathbb{R} \to M, \qquad \widetilde{\gamma} = \pi \circ \widetilde{g} : \mathbb{R} \to M$$

are gradient flow lines of the Kempf–Ness function $\Phi_{x_0} : M \to \mathbb{R}$. Theorem 4.3 also asserts that Φ_{x_0} is convex along geodesics. Since M is simply connected with non-positive sectional curvature, the function $\mathbb{R} \to \mathbb{R} : t \mapsto d_M(\gamma(t), \widetilde{\gamma}(t)) = |\eta(t)|$ is nonincreasing (see Lemma A.2). Hence there exists a sequence $t_\nu \to \infty$ such that the limits

$$\eta_\infty := \lim_{\nu\to\infty} \eta(t_\nu), \qquad u_\infty := \lim_{\nu\to\infty} u(t_\nu)$$

exist. Hence

$$\widetilde{x}_\infty := \lim_{t\to\infty} \widetilde{x}(t) = \lim_{t\to\infty} u(t)^{-1}\exp(-\mathbf{i}\eta(t))x(t) = u_\infty^{-1}\exp(-\mathbf{i}\eta_\infty)x_\infty.$$

This shows that x_∞ and $\widetilde{x}_\infty$ are critical points of the moment map squared belonging to the same G^c-orbit. Hence they belong to the same G-orbit by Theorem 6.3, and hence $|\mu(x_\infty)| = |\mu(\widetilde{x}_\infty)| \le |\mu(g_0^{-1}x_0)|$. This proves Theorem 6.4. □

Theorem 6.5 (Second Ness Uniqueness Theorem) *Let $x_0 \in X$ and*

$$m := \inf_{g\in\mathrm{G}^c} |\mu(gx_0)|.$$

Then

$$x, y \in \overline{\mathrm{G}^c(x_0)}, \quad |\mu(x)| = |\mu(y)| = m \qquad \Longrightarrow \qquad y \in \mathrm{G}(x).$$

For linear actions on projective space (and for $x, y \in \mathrm{G}^c(x_0)$) this is Theorem 6.2 (ii) in Ness [66]. For $m = 0$ it is Theorem 4.1 in Chen–Sun [23]. Theorem 6.5 shows that every element $x \in \overline{\mathrm{G}^c(x_0)}$ with

$$|\mu(x)| = m$$

is a critical point of the moment map squared and is the limit point of a negative gradient flow line of the moment map squared in $\mathrm{G}^c(x_0)$. In our proof of Theorem 6.5 we follow the argument of Chen–Sun.

Proof Let $x_0 \in X$, let $x : \mathbb{R} \to X$ be the unique solution of (3.2), and define

$$x_\infty := \lim_{t\to\infty} x(t).$$

Then

$$x_\infty \in \overline{\mathrm{G}^c(x_0)}, \qquad |\mu(x_\infty)| = m$$

by Theorem 6.4. Now let $x \in \overline{\mathrm{G}^c(x_0)}$ such that $|\mu(x)| = m$. We must prove that $x \in \mathrm{G}(x_\infty)$. To see this, choose a sequence $g_i \in \mathrm{G}^c$ such that

$$x = \lim_{i\to\infty} g_i^{-1} x_0$$

and define $y_i : \mathbb{R} \to X$ and $x_i \in X$ by

$$\dot{y}_i = -J L_{y_i} \mu(y_i), \qquad y_i(0) = g_i^{-1} x_0, \qquad x_i := \lim_{t\to\infty} y_i(t).$$

Then it follows from the estimate (3.4) in Theorem 3.3 that there exists a constant $c > 0$ such that, for i sufficiently large,

$$d(x_i, g_i^{-1} x_0) \le \int_0^\infty |\dot{y}_i(t)|\, dt \le c \left(|\mu(g_i^{-1} x_0)|^2 - m^2 \right)^{1-\alpha}.$$

Since

$$m = |\mu(x)| = \lim_{i\to\infty} |\mu(g_i^{-1} x_0)|,$$

this implies $x = \lim_{i\to\infty} x_i$. Moreover, we have

$$x_i \in \mathrm{G}(x_\infty)$$

for all i by Theorem 6.4. Hence $x \in \mathrm{G}(x_\infty)$ because the group orbit $\mathrm{G}(x_\infty)$ is compact. This proves Theorem 6.5. □

In general, the moment map squared is far from a Morse–Bott function and may have very complicated critical points. However, it follows from the Kirwan–Ness Inequality, the Moment Limit Theorem, and the Ness Uniqueness Theorem that the stable manifolds of this gradient flow exhibit a structure that resembles a stratification by stable manifolds associated to a Morse–Bott function. More precisely, let $x \in X$ be a critical point of the moment map squared, i.e. $L_x\mu(x) = 0$, and define the stable manifold of the critical set $\mathrm{G}(x)$ by

$$W^s(\mathrm{G}(x)) := \left\{ y_0 \in X \,\middle|\, \begin{array}{l} \text{the unique solution } y : \mathbb{R} \to X \text{ of} \\ \dot{y} = -JL_y\mu(y),\ y(0) = y_0 \text{ satisfies} \\ \lim_{t\to\infty} y(t) = ux \text{ for some } u \in \mathrm{G} \end{array} \right\}. \tag{6.10}$$

By Theorem (3.3) X is the union of these stable manifolds, and each stable manifold is a union of G^c-orbits by Theorems 6.4 and 6.5.

Corollary 6.6 *For $x \in \mathrm{Crit}(f)$, let $W^s(\mathrm{G}(x)) \subset X$ be the stable manifold in* (6.10). *Then the following holds.*

(i) $X = \bigcup_{x\in\mathrm{Crit}(f)} W^s(\mathrm{G}(x))$.
(ii) *Let $x \in \mathrm{Crit}(f)$ and $y_0 \in X$. Then $y_0 \in W^s(\mathrm{G}(x))$ if and only if*

$$x \in \overline{\mathrm{G}^c(y_0)}, \qquad |\mu(x)| = \inf_{g\in\mathrm{G}^c} |\mu(gy_0)| \tag{6.11}$$

(iii) *Let $x \in \mathrm{Crit}(f)$. Then $W^s(\mathrm{G}(x))$ is a union of G^c-orbits.*

Proof Part (i) follows directly from the Convergence Theorem 3.3. To prove part (ii), let $y_0 \in X$, let $y : \mathbb{R} \to X$ be the unique solution of the initial value problem $\dot{y} = -JL_y\mu(y)$ with $y(0) = y_0$, and define $y_\infty := \lim_{t\to\infty} y(t)$ (Theorem 3.3). Then, by Lemma 3.2 and Theorem 6.4, we have

$$y_\infty \in \overline{\mathrm{G}^c(y_0)}, \qquad |\mu(y_\infty)| = \inf_{g\in\mathrm{G}^c} |\mu(gy_0)|. \tag{6.12}$$

Moreover, it follows from the definition of $W^s(\mathrm{G}(x))$ in (6.10) that $y_0 \in W^s(\mathrm{G}(x))$ if and only if $y_\infty \in \mathrm{G}(x)$. Thus $y_0 \in W^s(\mathrm{G}(x))$ implies (6.11). Conversely, if y_0 satisfies (6.11) then it follows from (6.12) and the Second Ness Uniqueness Theorem 6.5 that $y_\infty \in \mathrm{G}(x)$ and hence $y_0 \in W^s(\mathrm{G}(x))$. This proves (ii). It follows from (ii) and the Moment Limit Theorem 6.4 that $W^s(\mathrm{G}(x))$ is a (possibly infinite) union of G^c-orbits. This proves (iii) and Corollary 6.6. □

The stratification in Corollary 6.6 was used by Kirwan [53] to prove that the canonical ring homomorphism (the **Kirwan homomorphism**)

$$\kappa : H^*_{\mathrm{G}}(X) \to H^*(X/\!\!/\mathrm{G})$$

from the equivariant cohomology of X to the cohomology of the Marsden–Weinstein quotient $X/\!\!/\mathrm{G}$ with rational coefficients is surjective.

Theorem 6.7 (General Moment-Weight Inequality) *For every element $x \in X$, every $\xi \in \mathfrak{g} \setminus \{0\}$, and every $g \in \mathrm{G}^c$,*

$$\frac{-w_\mu(x,\xi)}{|\xi|} \leq |\mu(gx)|. \tag{6.13}$$

We give two proofs of Theorem 6.7. The first proof is due to Mumford [63] and Ness [66, Lemma 3.1] and is based on Theorem D.4. The second proof is due to Chen [15, 16]. His methods were developed in the infinite-dimensional setting of K-stability for Kähler–Einstein metrics (Tian [75], Donaldson [32–34]). Chen's infinite-dimensional argument carries over to the finite-dimensional setting.

Proof 1 We first prove the inequality (6.13) for $g = 1$. Choose elements $x_0 \in X$ and $\xi \in \mathfrak{g} \setminus \{0\}$. Define $x : \mathbb{R} \to X$ by

$$x(t) := \exp(\mathbf{i}t\xi)x_0$$

as in Lemma 5.4. Then the function $t \mapsto \langle \mu(x(t)), \xi \rangle$ is nondecreasing. Hence

$$\langle \mu(x_0), \xi \rangle = \langle \mu(x(0)), \xi \rangle \leq \lim_{t\to\infty} \langle \mu(x(t)), \xi \rangle = w_\mu(x_0, \xi).$$

Hence, by the Cauchy–Schwarz inequality,

$$-w_\mu(x_0,\xi) \leq -\langle \mu(x_0), \xi \rangle \leq |\mu(x_0)||\xi|$$

and this implies (6.13) with $g = 1$.

Now let $x \in X$, $\xi \in \mathfrak{g} \setminus \{0\}$, and $g \in \mathrm{G}^c$. Then

$$\zeta := g\xi\, g^{-1} \in \mathscr{T}^c.$$

Hence it follows from Theorem D.4 that there exists an element $p \in \mathrm{P}(\zeta)$ such that $p\zeta p^{-1} \in \mathfrak{g} \setminus \{0\}$. Thus it follows from part (ii) of Theorem 5.3 that

$$w_\mu(gx, p\zeta p^{-1}) = w_\mu(gx, \zeta).$$

Now apply the first step of the proof to the pair $(gx, p\zeta p^{-1}) \in X \times (\mathfrak{g} \setminus \{0\})$. Then

$$\begin{aligned}|\mu(gx)| &\geq \frac{-w_\mu(gx, p\zeta p^{-1})}{|p\zeta p^{-1}|} \\ &= \frac{-w_\mu(gx, \zeta)}{\sqrt{|\mathrm{Re}(\zeta)|^2 - |\mathrm{Im}(\zeta)|^2}} \\ &= \frac{-w_\mu(x, \xi)}{|\xi|}.\end{aligned}$$

Here the last step follows from part (i) of Theorem 5.3, Lemma 5.7, and the fact that $\zeta = g\xi g^{-1}$. This completes the first proof of Theorem 6.7. □

Proof 2 Let $x \in X$, $\xi \in \mathfrak{g} \setminus \{0\}$, and $g \in \mathrm{G}^c$. For $t \geq 0$ choose $\eta(t) \in \mathfrak{g}$ and $u(t) \in \mathrm{G}$ such that

$$\exp(\mathbf{i}\eta(t))u(t) = \exp(\mathbf{i}t\xi)g^{-1}. \tag{6.14}$$

We prove that

$$\lim_{t\to\infty} \frac{\eta(t)}{|\eta(t)|} = \frac{\xi}{|\xi|}. \tag{6.15}$$

To see this, note that

$$\exp(-\mathbf{i}\eta(t))\exp(\mathbf{i}t\xi)g^{-1} \in \mathrm{G}.$$

Hence it follows from Lemma C.2 that there exists a constant $c > 0$ such that

$$|t\xi - \eta(t)| \leq c \qquad \text{for all } t \geq 0.$$

Hence

$$\begin{aligned}\left|\frac{\xi}{|\xi|} - \frac{\eta(t)}{|\eta(t)|}\right| &\leq \frac{|t\xi - \eta(t)|}{t|\xi|} + |\eta(t)|\left|\frac{1}{t|\xi|} - \frac{1}{|\eta(t)|}\right| \\ &= \frac{|t\xi - \eta(t)|}{t|\xi|} + \frac{||t\xi| - |\eta(t)||}{t|\xi|} \\ &\leq \frac{2c}{t|\xi|}.\end{aligned}$$

This proves (6.15). Now abbreviate $u := u(t)$ and $\eta := \eta(t)$. Then, by (6.14),

$$\exp(\mathbf{i}u^{-1}\eta u)g = u^{-1}\exp(\mathbf{i}t\xi).$$

Moreover, by Eq. (2.7), the function

$$s \mapsto \langle \mu(\exp(\mathbf{i}su^{-1}\eta u)gx), u^{-1}\eta u\rangle$$

is nondecreasing. Hence

$$\begin{aligned}
-|\mu(gx)| &\le |\eta|^{-1}\langle \mu(gx), u^{-1}\eta u\rangle \\
&\le |\eta|^{-1}\langle \mu(\exp(\mathbf{i}u^{-1}\eta u)gx), u^{-1}\eta u\rangle \\
&= |\eta|^{-1}\langle \mu(u^{-1}\exp(\mathbf{i}t\xi)x), u^{-1}\eta u\rangle \\
&= |\eta|^{-1}\langle \mu(\exp(\mathbf{i}t\xi)x), \eta\rangle \\
&= |\xi|^{-1}\langle \mu(\exp(\mathbf{i}t\xi)x), \xi\rangle + \langle \mu(\exp(\mathbf{i}t\xi)x), |\eta|^{-1}\eta - |\xi|^{-1}\xi\rangle.
\end{aligned}$$

Take the limit $t \to \infty$ and use (6.15) to obtain

$$-|\mu(gx)| \le \lim_{t\to\infty} \frac{\langle \mu(\exp(\mathbf{i}t\xi)x), \xi\rangle}{|\xi|} = \frac{w_\mu(x,\xi)}{|\xi|}.$$

This completes the second proof of Theorem 6.7. □

Here is the geometric picture behind Chen's proof in [15] of the moment-weight inequality (proof 2 above). The homogeneous space $M = \mathrm{G}^c/\mathrm{G}$ is a complete, connected, simply connected Riemannian manifold with nonpositive sectional curvature (Appendix C). The curve

$$\gamma(t) := \pi(\exp(-\mathbf{i}t\xi))$$

is a geodesic through the point $p_0 := \pi(\mathbb{1})$. The formula (6.14) asserts that, for each t, the curve

$$\beta_t(s) := \pi(g^{-1}\exp(-\mathbf{i}su(t)^{-1}\eta(t)u(t)))$$

is the unique geodesic connecting $p_1 := \pi(g^{-1}) = \beta_t(0)$ to $\gamma(t) = \beta_t(1)$ (see Fig. 6.1). Equation (6.15) asserts that the angle between the geodesics β_t and γ

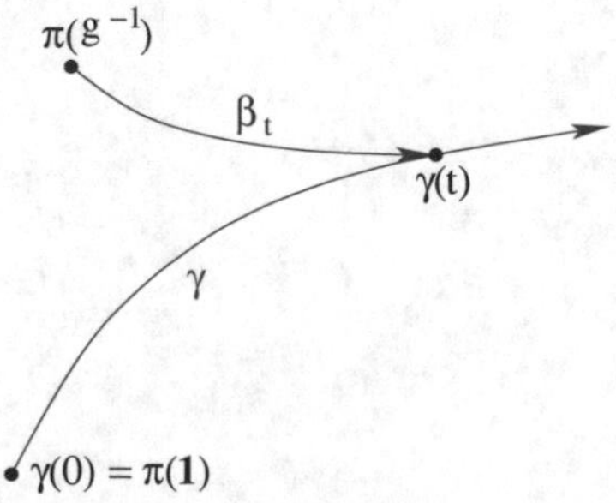

Fig. 6.1 Chen's proof of the moment-weight inequality

at the point $\gamma(t)$ where they meet tends to zero as t tends to infinity. (This is closely related to the sphere at infinity discussed in Donaldson's paper [36].) Now the main point of Chen's proof is a comparison between the slope of the Kempf–Ness function $\Phi_x : M \to \mathbb{R}$ along the geodesics γ and β_t at the point $\gamma(t)$.

Chapter 7
Stability in Symplectic Geometry

This chapter introduces the μ-stability conditions for elements of X and characterizes them in terms of the properties of the Kempf–Ness function Φ_x. This is the content of the generalized Kempf–Ness Theorem 7.3.

Definition 7.1 An element $x \in X$ is called
μ-**unstable** iff $\overline{\mathrm{G}^c(x)} \cap \mu^{-1}(0) = \emptyset$,
μ-**semistable** iff $\overline{\mathrm{G}^c(x)} \cap \mu^{-1}(0) \neq \emptyset$,
μ-**polystable** iff $\mathrm{G}^c(x) \cap \mu^{-1}(0) \neq \emptyset$,
μ-**stable** iff $\mathrm{G}^c(x) \cap \mu^{-1}(0) \neq \emptyset$ and $\mathrm{G}^c_x := \{g \in \mathrm{G}^c \mid gx = x\}$ is discrete.
Denote the sets of μ-semistable, μ-polystable, and μ-stable points by

$$\begin{aligned} X^{\mathrm{ss}} &:= \{x \in X \mid x \text{ is } \mu\text{-semistable}\}, \\ X^{\mathrm{ps}} &:= \{x \in X \mid x \text{ is } \mu\text{-polystable}\}, \\ X^{\mathrm{s}} &:= \{x \in X \mid x \text{ is } \mu\text{-stable}\}. \end{aligned} \tag{7.1}$$

With this terminology Theorem 6.5 asserts for $m = 0$ that the closure of each μ-semistable G^c-orbit in X contains a unique μ-polystable G^c-orbit. This gives rise to a bijection $X^{\mathrm{ps}}/\mathrm{G}^c \cong X^{\mathrm{ss}}/{\sim}$, where two μ-semistable points $x, x' \in X^{\mathrm{ss}}$ are equivalent iff the closures of their G^c-orbits contain the same μ-polystable orbit, i.e.

$$x \sim x' \quad \stackrel{\text{def}}{\Longleftrightarrow} \quad \overline{\mathrm{G}^c(x)} \cap \overline{\mathrm{G}^c(x')} \cap X^{\mathrm{ss}} \neq \emptyset.$$

The next theorem characterizes the μ-stability conditions in terms of the negative gradient flow lines of the square of the moment map.

Theorem 7.2 (μ-Stability Theorem) *Let $x_0 \in X$ and let $x : \mathbb{R} \to X$ be the solution of* (3.2). *Define $x_\infty := \lim_{t\to\infty} x(t)$. Then the following holds.*

© The Author(s), under exclusive license to Springer Nature Switzerland AG 2021
V. Georgoulas et al., *The Moment-Weight Inequality and the Hilbert–Mumford Criterion*, Lecture Notes in Mathematics 2297,
https://doi.org/10.1007/978-3-030-89300-2_7

(i) $x_0 \in X^{\mathrm{ss}}$ *if and only if* $\mu(x_\infty) = 0$.
(ii) $x_0 \in X^{\mathrm{ps}}$ *if and only if* $\mu(x_\infty) = 0$ *and* $x_\infty \in \mathrm{G}^c(x_0)$.
(iii) $x_0 \in X^{\mathrm{s}}$ *if and only if the isotropy subgroup* G_{x_∞} *is discrete.*

Moreover, X^{ss} *and* X^{s} *are open subsets of* X.

Proof By definition $x_0 \in X^{\mathrm{ss}}$ if and only if $\inf_{g\in \mathrm{G}^c} |\mu(gx_0)| = 0$. Hence assertion (i) follows from Theorem 6.4.

We prove that X^{ss} is open. By the Lojasiewicz gradient inequality (3.5) for the function $f = \frac{1}{2}|\mu|^2$ there exists a constant $\delta > 0$ such that every $x \in X$ satisfies

$$f(x) < \delta, \quad \nabla f(x) = 0 \qquad \Longrightarrow \qquad f(x) = 0.$$

Hence it follows from (i) that

$$U := \left\{x \in X \mid \tfrac{1}{2}|\mu(x)|^2 < \delta\right\} \subset X^{\mathrm{ss}}.$$

Now let $x_0 \in X^{\mathrm{ss}}$. Then there is a $g_0 \in \mathrm{G}^c$ such that $g_0x_0 \in U$. Hence there exists a neighborhood $V \subset X$ of x_0 such that $g_0V \subset U$, and thus $V \subset X^{\mathrm{ss}}$. This shows that X^{ss} is an open subset of X.

We prove (ii). If $\mu(x_\infty) = 0$ and $x_\infty \in \mathrm{G}^c(x_0)$ then $\mathrm{G}^c(x_0) \cap \mu^{-1}(0) \neq \emptyset$ and hence $x_0 \in X^{\mathrm{ps}}$. Conversely, suppose that $x_0 \in X^{\mathrm{ps}}$. Then there exists an element $g_0 \in \mathrm{G}^c$ such that $\mu(g_0x_0) = 0$. In particular, g_0x_0 is a critical point of $f = \frac{1}{2}|\mu|^2$ and thus the negative gradient flow line of f with the initial value g_0x_0 is constant. Hence it follows from Theorem 6.4 that $g_0x_0 \in \mathrm{G}(x_\infty)$. Hence $x_\infty \in \mathrm{G}^c(x_0)$ and $\mu(x_\infty) = 0$. This proves (ii).

We prove that X^{s} is open. By Lemma 2.2 the set

$$Z^{\mathrm{s}} := \{x \in X \mid \mu(x) = 0,\ \ker L_x = 0\}$$

is a smooth submanifold of X with tangent spaces $T_xZ^{\mathrm{s}} = \ker d\mu(x)$. It follows also from Lemma 2.2 and the definition of μ-stability that $X^{\mathrm{s}} = \mathrm{G}^cZ^{\mathrm{s}}$. Next we prove that there exists an open set $U^{\mathrm{s}} \subset X$ such that

$$Z^{\mathrm{s}} \subset U^{\mathrm{s}} \subset X^{\mathrm{s}}.$$

Define the map $\psi : Z^{\mathrm{s}} \times \mathfrak{g} \to X$ by

$$\psi(x, \eta) := \exp(\mathbf{i}\eta)x$$

for $x \in X$ and $\eta \in \mathfrak{g}$. Its derivative at an element $(x, 0) \in Z^{\mathrm{s}} \times \{0\}$ is the linear map $d\psi(x, 0) : \ker d\mu(x) \times \mathfrak{g} \to T_xX$ given by

$$d\psi(x, 0)(\widehat{x}, \widehat{\eta}) = \widehat{x} + JL_x\widehat{\eta}$$

for $\widehat{x} \in T_x X$ and $\widehat{\eta} \in \mathfrak{g}$. We prove that this map is bijective for every $x \in Z^{\mathrm{s}}$. If $\widehat{x} \in \ker d\mu(x)$ and $\widehat{\eta} \in \mathfrak{g}$ satisfy $\widehat{x} + JL_x\widehat{\eta} = 0$, then $0 = d\mu(x)JL_x\widehat{\eta} = L_x{}^*L_x\widehat{\eta}$ by (2.6), hence $\widehat{\eta} = 0$, and hence $\widehat{x} = 0$. Moreover, the map $d\mu(x)$ is surjective and so $\dim \ker d\mu(x) = \dim X - \dim \mathfrak{g}$. This shows that $d\psi(x, 0)$ is bijective for each $x \in Z^{\mathrm{s}}$. Hence ψ restricts to a diffeomorphism from an open neighborhood of $Z^{\mathrm{s}} \times \{0\}$ in $Z^{\mathrm{s}} \times \mathfrak{g}$ to an open neighborhood $U^{\mathrm{s}} \subset X$ of Z^{s}. Since $Z^{\mathrm{s}} \subset U^{\mathrm{s}} \subset X^{\mathrm{s}}$, it follows that $X^{\mathrm{s}} = \bigcup_{g \in \mathrm{G}^c} gU^{\mathrm{s}}$ is an open subset of X.

We prove (iii). Let $x_0 \in X$, let $x : \mathbb{R} \to X$ be the unique solution of (3.2), and define $x_\infty := \lim_{t\to\infty} x(t)$. Assume first that $x_0 \in X^{\mathrm{s}}$. Then it follows from (ii) and the definition of μ-stability that

$$\mu(x_\infty) = 0, \qquad x_\infty \in \mathrm{G}^c(x_0), \qquad \ker L^c_{x_0} = 0.$$

Hence $\ker L^c_{x_\infty} = 0$, thus $\ker L_{x_\infty} = 0$ by Lemma 2.2, and so G_{x_∞} is discrete. Conversely, assume G_{x_∞} is discrete and so $\ker L_{x_\infty} = 0$. Since $L_{x_\infty}\mu(x_\infty) = 0$ by Theorem 3.3, this implies $\mu(x_\infty) = 0$. Thus x_∞ is μ-stable. Since X^{s} is open, $x(t)$ is μ-stable for t sufficiently large and, since $x(t) \in \mathrm{G}^c(x_0)$ for all t by Lemma 3.2, x_0 is μ-stable. This proves (iii) and Theorem 7.2. □

The generalized Kempf–Ness theorem characterizes the stability condition of an element $x \in X$ in terms of the properties of the *Kempf–Ness function* Φ_x. We will derive the original Kempf–Ness theorem [52] as a corollary in Theorem 8.5.

Theorem 7.3 (Generalized Kempf–Ness Theorem) *Let $x \in X$ and denote by $\Phi_x : M \to \mathbb{R}$ the Kempf–Ness function of x. Then the following holds.*

(i) *x is μ-unstable if and only if Φ_x is unbounded below.*
(ii) *x is μ-semistable if and only if Φ_x is bounded below.*
(iii) *x is μ-polystable if and only if Φ_x has a critical point.*
(iv) *x is μ-stable if and only if Φ_x is bounded below and proper.*

Proof Fix a point $x_0 \in X$ and denote by $\Phi : M \to \mathbb{R}$ the Kempf–Ness function of x_0. Throughout the proof $x : \mathbb{R} \to X$ denotes the solution of (3.2), $g : \mathbb{R} \to \mathrm{G}^c$ denotes the solution of (3.3) so that $x(t) = g(t)^{-1}x_0$ for all t, and $\gamma : \mathbb{R} \to M$ denotes the composition of g with the projection from G^c to M. The limit $x_\infty := \lim_{t\to\infty} x(t)$ exists by Theorems 3.3, and 6.4 asserts that $|\mu(x_\infty)| = \inf_{g\in\mathrm{G}^c} |\mu(gx_0)|$.

We prove necessity in (i). Assume x_0 is μ-unstable. Then $\mu(x_\infty) \neq 0$. By definition of the Kempf–Ness function

$$\frac{d}{dt}(\Phi \circ \gamma) = -\langle \mu(g^{-1}x_0), \mathrm{Im}(g^{-1}\dot{g})\rangle = -|\mu(x)|^2 \leq -|\mu(x_\infty)|^2.$$

This implies $\Phi(\gamma(t)) \leq -t|\mu(x_\infty)|^2$ for all $t \geq 0$. Thus Φ is unbounded below.

We prove necessity in (ii). Assume x_0 is μ-semistable. Then $\mu(x_\infty) = 0$. By the Lojasiewicz gradient inequality for $f = \frac{1}{2}|\mu|^2$ in (3.5) there exist positive constants t_0 and c, and a constant $1/2 < \alpha < 1$, such that

$$|\mu(x)|^2 = 2|f(x)| \leq 2|f(x)|^\alpha \leq c|\nabla f(x)| = c|JL_x\mu(x)| = c|\dot{x}|$$

for $t \geq t_0$. By Theorem 3.3 the function $t \mapsto |\dot{x}(t)|$ is integrable over the positive real axis and hence, so is the function $t \mapsto |\mu(x(t))|^2 = -\frac{d}{dt}(\Phi\circ\gamma)(t)$. This implies that the limit $a := \lim_{t\to\infty} \Phi(\gamma(t))$ exists in $\mathbb{R}$. Hence $a = \inf_M \Phi$ by Theorem 4.3 and so Φ is bounded below.

Thus we have proved that the conditions on the Kempf–Ness function are necessary in (i) and (ii). Since necessity in (i) is equivalent to sufficiency in (ii) and vice versa, this proves (i) and (ii). Assertion (iii) follows from the fact that $\pi(g)$ is a critical point of Φ if and only if $\mu(g^{-1}x_0) = 0$.

We prove (iv). Assume first that Φ is bounded below and proper. Then x_0 is μ-semistable by part (ii). Moreover, by definition of the Kempf–Ness function Φ and the negative gradient flow line γ, we have

$$\Phi(\gamma(0)) = 0, \qquad -\infty < a := \inf_M \Phi \leq 0,$$

and the function $\Phi \circ \gamma : \mathbb{R} \to \mathbb{R}$ is nonincreasing. Since Φ is proper the set $\Phi^{-1}([a, 0])$ is compact and contains $\gamma(t)$ for every $t \geq 0$. Hence there exists a sequence $t_i \to \infty$ such that $\gamma(t_i)$ converges. Hence Φ has a critical point. By (iii) this implies that x_0 is μ-polystable. Choose $g_0 \in \mathrm{G}^c$ such that $\mu(g_0x_0) = 0$. Assume by contradiction that x_0 is not μ-stable. Then $\ker L^c_{x_0} \neq 0$, hence $\ker L^c_{g_0x_0} \neq 0$, and hence $\ker L_{g_0x_0} \neq 0$ by Lemma 2.2. Choose $\eta \in \mathfrak{g} \setminus \{0\}$ such that $L_{g_0x_0}\eta = 0$. Then $\exp(\mathbf{i}s\eta)g_0x_0 = g_0x_0$ and so $\mu(\exp(\mathbf{i}s\eta)g_0x_0) = 0$. Thus the curve $\beta(s) := \pi(g_0^{-1}\exp(-\mathbf{i}s\eta))$ consists of critical points of Φ and hence, by part (iii) of Theorem 4.3, $\Phi(\beta(s)) = a$ for all s. Thus $\Phi^{-1}(a)$ is not compact in contradiction to the assumption that Φ is proper. Thus x_0 is μ-stable.

Conversely assume that x_0 is μ-stable. Then Φ is bounded below by part (ii) and we must prove that Φ is proper. For this it suffices to assume that $\mu(x_0) = 0$, by part (vii) of Theorem 4.3. We prove first that $p_0 := \pi(\mathbb{1}) \in M$ is the unique point at which Φ attains its minimum $\min_M \Phi = 0$. To see this, let $p = \pi(g) \in M \setminus \{p_0\}$. Choose $\eta \in \mathfrak{g}$ and $u \in \mathrm{G}$ such that $g = \exp(\mathbf{i}\eta)u$. Then $\eta \neq 0$ and define

$$y(t) := \exp(\mathbf{i}t\eta)x_0, \qquad \phi(t) := \Phi(\pi(\exp(-\mathbf{i}t\eta))).$$

Then the function $\phi : \mathbb{R} \to \mathbb{R}$ is convex and

$$\dot{\phi}(t) = \langle \mu(y(t)), \eta \rangle, \qquad \ddot{\phi}(t) = |L_{y(t)}\eta|^2.$$

Since $y(0) = x_0$ we obtain $\phi(0) = 0$, $\dot{\phi}(0) = 0$, and $\ddot{\phi}(0) = |L_{x_0}\eta|^2 > 0$. This implies $\phi(t) > 0$ for every $t \in \mathbb{R} \setminus 0$ and, in particular, $\Phi(p) = \phi(-1) > 0$.

For every $r > 0$ denote by $B_r \subset M$ the ball of radius r centered at p_0. Define the number $\delta := \inf_{\partial B_1} \Phi > 0$. Then it follows from convexity that

$$d(p_0, p) \geq 1 \qquad \Longrightarrow \qquad \Phi(p) \geq \delta d_M(p_0, p)$$

for all $p \in M$. Hence, for every $c > 0$,

$$c \geq \delta \qquad \Longrightarrow \qquad \Phi^{-1}([0, c]) \subset B_{c/\delta}.$$

This shows that, for every $c > 0$, the set $\Phi^{-1}([0, c])$ is closed and bounded, and hence compact because M is complete. This proves (iv) and Theorem 7.3. □

A central result in geometric invariant theory is the Hilbert–Mumford criterion, which characterizes the μ-stability conditions in terms of the μ-weights introduced in Chap. 5. The necessity of these conditions follows directly from the general moment-weight inequality (6.13) in Theorem 6.7.

Theorem 7.4 (Stability and Weights) *Let $x_0 \in X$. Then the following holds.*

(i) *If x_0 is μ-semistable and $\zeta \in \mathscr{T}^c$ then $w_\mu(x_0, \zeta) \geq 0$.*
(ii) *If x_0 is μ-polystable and $\zeta \in \mathscr{T}^c$ then $w_\mu(x_0, \zeta) \geq 0$ and*

$$w_\mu(x_0, \zeta) = 0 \qquad \Longleftrightarrow \qquad \lim_{t\to\infty} \exp(\mathbf{i}t\zeta)x_0 \in \mathrm{G}^c(x_0).$$

(iii) *If x_0 is μ-stable and $\zeta \in \mathscr{T}^c$ then $w_\mu(x_0, \zeta) > 0$.*

Proof See page 56. □

The Hilbert–Mumford criterion asserts that the necessary conditions for μ-semistability, μ-polystability, and μ-stability in Theorem 7.4 are in fact necessary and sufficient. This is proved in Chap. 12 below.

Lemma 7.5 *Let $x_0 \in X$ such that $\mu(x_0) = 0$ and let $\xi \in \mathfrak{g} \setminus \{0\}$. Then the following are equivalent.*

(i) $w_\mu(x_0, \xi) = 0$.
(ii) $L_{x_0}\xi = 0$.
(iii) $\lim_{t\to\infty} \exp(\mathbf{i}t\xi)x_0 \in \mathrm{G}^c(x_0)$.

Proof Assume (i) and define $x(t) := \exp(\mathbf{i}t\xi)x_0$. Then, by (i) and (2.7),

$$\langle\mu(x(0)), \xi\rangle = 0, \qquad \lim_{t\to\infty} \langle\mu(x(t)), \xi\rangle = 0, \qquad \frac{d}{dt}\langle\mu(x(t)), \xi\rangle = |L_{x(t)}\xi|^2$$

for all t. Hence $L_{x(t)}\xi = 0$ for all t and this shows that (i) implies (ii). That (ii) implies (iii) follows from the fact that $\exp(\mathbf{i}t\xi)x_0 = x_0$ for $t \in \mathbb{R}$ and $\xi \in \ker L_{x_0}$. Now assume (iii) and define

$$x^+ := \lim_{t\to\infty} \exp(\mathbf{i}t\xi)x_0.$$

By (iii) there exists an element $g \in \mathrm{G}^c$ such that $x^+ = gx_0$. Since $L_{x^+}\xi = 0$ by Lemma 5.4, we have

$$\begin{aligned} w_\mu(x_0, \xi) &= \langle \mu(x^+), \xi \rangle = w_\mu(x^+, \xi) \\ &= w_\mu(gx_0, \xi) = w_\mu(x_0, g^{-1}\xi g) \\ &= \langle \mu(x_0), \mathrm{Re}(g^{-1}\xi g) \rangle = 0. \end{aligned}$$

Here we have used Theorem 5.3 and the fact that $\exp(\mathbf{i}tg^{-1}\xi g)x_0 = x_0$ for all t. Thus (iii) implies (i) and this proves Lemma 7.5. □

Lemma 7.6 *Let $x_0 \in X$ such that $\mu(x_0) = 0$ and let $\zeta \in \mathscr{T}^c$. Then*

$$w_\mu(x_0, \zeta) = 0 \qquad \iff \qquad x^+ := \lim_{t\to\infty} \exp(\mathbf{i}t\zeta)x_0 \in \mathrm{G}^c(x_0).$$

Proof By Theorem D.4 there exists a $p \in \mathrm{P}(\zeta)$ such that $\xi := p\zeta p^{-1} \in \mathfrak{g}$. Then the limit

$$p^+ := \lim_{t\to\infty} \exp(\mathbf{i}t\zeta)p\exp(-\mathbf{i}t\zeta)$$

exists in G^c. It satisfies

$$\begin{aligned} x^+ &= \lim_{t\to\infty} \exp(\mathbf{i}t\zeta)x_0 \\ &= \lim_{t\to\infty} \big(\exp(\mathbf{i}t\zeta)p\exp(-\mathbf{i}t\zeta)\big) \lim_{t\to\infty} \exp(\mathbf{i}t\zeta)p^{-1}x_0 \\ &= p^+p^{-1} \lim_{t\to\infty} p\exp(\mathbf{i}t\zeta)p^{-1}x_0 \\ &= p^+p^{-1} \lim_{t\to\infty} \exp(\mathbf{i}t\xi)x_0. \end{aligned}$$

This shows that $x^+ \in \mathrm{G}^c(x_0)$ if and only if $\lim_{t\to\infty} \exp(\mathbf{i}t\xi)x_0 \in \mathrm{G}^c(x_0)$. Moreover, we have $w_\mu(x_0, \xi) = w_\mu(x_0, \zeta)$ by part (ii) of Theorem 5.3. Hence it follows from Lemma 7.5 that $w_\mu(x_0, \zeta) = 0$ if and only if $x^+ \in \mathrm{G}^c(x_0)$, and this proves Lemma 7.6. □

Proof of Theorem 7.4 We prove part (i) by an indirect argument. Suppose that there exists a toral generator $\zeta \in \mathscr{T}^c$ such that $w_\mu(x_0, \zeta) < 0$. Then, by Theorem D.4 and part (ii) of Theorem 5.3, there exists a $\xi \in \mathfrak{g} \setminus \{0\}$ such that $w_\mu(x_0, \xi) < 0$. Hence $\inf_{g\in\mathrm{G}^c} |\mu(gx_0)| > 0$ by Theorem 6.7 and so x_0 is μ-unstable. This proves part (i).

We prove part (ii). Suppose x_0 is μ-polystable and fix an element $\zeta \in \mathscr{T}^c$. Then $w_\mu(x_0, \zeta) \geq 0$ by part (i). Now choose $g \in \mathrm{G}^c$ such that $\mu(gx_0) = 0$ and define $x^+ := \lim_{t\to\infty} \exp(\mathbf{i}t\zeta)x_0$. Then

$$\lim_{t\to\infty} \exp(\mathbf{i}tg\zeta g^{-1})gx_0 = gx^+.$$

Hence Lemma 7.6 asserts that $w_\mu(gx_0, g\zeta g^{-1}) = 0$ if and only if $gx^+ \in \mathrm{G}^c(gx_0)$. Moreover, we have $w_\mu(gx_0, g\zeta g^{-1}) = w_\mu(x_0, \zeta)$ by part (i) of Theorem 5.3, and this shows that $w_\mu(x_0, \zeta) = 0$ if and only if $x^+ \in \mathrm{G}^c(x_0)$. This proves part (ii).

We prove part (iii). Assume that x_0 is μ-stable. Then $w_\mu(x_0, \zeta) \geq 0$ for all $\zeta \in \mathscr{T}^c$ by part (i). Moreover, there exists an element $g \in \mathrm{G}^c$ such that

$$\mu(gx_0) = 0, \qquad \ker L_{gx_0} = \{0\}. \tag{7.2}$$

Suppose, by contradiction, that there exists a $\zeta \in \mathscr{T}^c$ such that $w_\mu(x_0, \zeta) = 0$. Then $w_\mu(gx_0, g\zeta g^{-1}) = 0$ by part (i) of Theorem 5.3. Hence it follows from Theorem D.4 and part (ii) of Theorem 5.3, that there exists an element $\xi \in \mathfrak{g} \setminus \{0\}$ such that $w_\mu(gx_0, \xi) = 0$. Hence $L_{gx_0}\xi = 0$ by Lemma 7.5, in contradiction to (7.2). This proves part (iii) and Theorem 7.4. □

Remark 7.7

(i) In Lemma 7.5 the hypothesis $\mu(x_0) = 0$ cannot be replaced by the hypothesis that x_0 is μ-polystable. For example, consider the diagonal action of $\mathrm{G} := \mathrm{SO}(3)$ on $X := S^2 \times S^2$, where $S^2 \subset \mathbb{R}^3$ is the unit sphere with its standard symplectic form. Let

$$x = (-\xi, \eta) \in S^2 \times S^2$$

be a pair of distinct non-antipodal points on S^2, so that

$$\eta \neq \pm\xi.$$

Then x is μ-polystable, $\mu(x) \neq 0$, and the isotropy subgroup of x in SO(3) is trivial. Identify the Lie algebra $\mathfrak{g} := \mathfrak{so}(3)$ with $\mathbb{R}^3$ so that $\xi \in S^2 \subset \mathfrak{g}$. Then

$$\lim_{t\to\infty} \exp(\mathbf{i}t\xi)x = (-\xi, \xi) \in \mathrm{G}^c(x), \qquad w_\mu(x, \xi) = 0, \qquad L_x\xi \neq 0. \tag{7.3}$$

(ii) In Lemma 7.5 the element $\xi \in \mathfrak{g} \setminus \{0\}$ cannot be replaced by $\zeta \in \mathscr{T}^c$. For example, let $(x, \xi) \in X \times (\mathfrak{g} \setminus \{0\})$ be as in part (i) so that x is μ-polystable and (7.3) holds. Choose an element $g \in \mathrm{G}^c$ such that $\mu(gx) = 0$ and define

$$x_0 := gx, \qquad \zeta := g\xi g^{-1} \in \mathscr{T}^c.$$

Then, by part (i) of Theorem 5.3, we have

$$\mu(x_0) = 0, \qquad w_\mu(x_0, \zeta) = 0, \qquad L^c_{x_0}\zeta \neq 0. \tag{7.4}$$

(iii) Let $x_0 \in X$ be μ-polystable and fix an element $\zeta \in \mathscr{T}^c$. Then

$$L_{x_0}^c \zeta = 0 \qquad \Longrightarrow \qquad w_\mu(x_0, \zeta) = 0. \tag{7.5}$$

To see this, assume that $L_{x_0}^c \zeta = 0$ and choose $g \in \mathrm{G}^c$ such that $\mu(gx_0) = 0$. Then $L_{gx_0}^c(g\zeta g^{-1}) = 0$ and hence it follows from part (i) of Theorem 5.3 that

$$w_\mu(x_0, \zeta) = w_\mu(gx_0, g\zeta g^{-1}) = \langle \mu(gx_0), \mathrm{Re}(g\zeta g^{-1}) \rangle = 0.$$

This proves (7.5). By part (ii) the converse does not hold, i.e. $w_\mu(x_0, \zeta) = 0$ does not imply $L_{x_0}^c \zeta = 0$, even in the case $\mu(x_0) = 0$.

(iv) The hypothesis that x_0 is μ-polystable cannot be removed in (7.5). For example, consider the standard action of $\mathrm{G} = \mathrm{SO}(3)$ on $X = S^2$, fix an element $x_0 \in S^2$, and choose $\xi := x_0 \in S^2 \subset \mathfrak{g}$. Then x_0 is μ-unstable and $L_{x_0} \xi = 0$, however, the Mumford weights are $w_\mu(x_0, \pm\xi) = \pm 1$.

Chapter 8
Stability in Algebraic Geometry

This chapter examines linear group actions on projective space, which is the classical setting of geometric invariant theory [51, 52, 63, 65, 66].

Let V be a finite-dimensional complex vector space, let $\mathrm{G} \subset \mathrm{U}(n)$ be a compact Lie group equipped with a representation $\mathrm{G} \to \mathrm{GL}(V)$, and let $\mathrm{G}^c \subset \mathrm{GL}(n, \mathbb{C})$ be its complexification and

$$\mathrm{G}^c \times V \to V : (g, v) \mapsto gv$$

be the complexified representation.

Definition 8.1 A nonzero vector $v \in V$ is called

unstable iff $0 \in \overline{\mathrm{G}^c(v)}$,
semistable iff $0 \notin \overline{\mathrm{G}^c(v)}$,
polystable iff $\mathrm{G}^c(v) = \overline{\mathrm{G}^c(v)}$,
stable iff $\mathrm{G}^c(v) = \overline{\mathrm{G}^c(v)}$ and the isotropy subgroup G^c_v is discrete.

The linear action of G^c on V induces an action on projective space $\mathbb{P}(V)$. Fix a G-invariant Hermitian structure on V and denote by $\langle \cdot, \cdot \rangle$ the associated real inner product on V. Choose a scaling factor[1]

$$\hbar > 0$$

[1] Think of $\hbar$ as Planck's constant. If the symplectic form ω is replaced by a positive real multiple $c\,\omega$ and $\hbar$ is replaced by $c\,\hbar$ all the formulas which follow remain correct. Our choice of $\hbar$ is consistent with the physics notation in that $\omega, \mu, \hbar$ have the units of action and the Hamiltonian $H_\xi = \langle \mu, \xi \rangle$ has the units of energy.

© The Author(s), under exclusive license to Springer Nature Switzerland AG 2021
V. Georgoulas et al., *The Moment-Weight Inequality and the Hilbert–Mumford Criterion*, Lecture Notes in Mathematics 2297,
https://doi.org/10.1007/978-3-030-89300-2_8

and restrict the symplectic structure to the sphere of radius

$$r := \sqrt{2\hbar}.$$

It is S^1-invariant and descends to $\mathbb{P}(V)$. (This is $2\hbar$ times the Fubini–Study form. Its integral over the positive generator of $\pi_2(\mathbb{P}(V))$ is $2\pi\hbar$.)

Lemma 8.2 *A moment map for the action of* G *on* $\mathbb{P}(V)$ *is given by*

$$\langle \mu(x), \xi \rangle = \hbar \frac{\langle v, \mathbf{i}\xi v \rangle}{|v|^2}, \qquad x := [v] \in \mathbb{P}(V). \tag{8.1}$$

Proof Define the map $\mu : \mathbb{P}(V) \to \mathfrak{g}^*$ by (8.1). Fix an element $x \in \mathbb{P}(V)$ and a tangent vectors $\widehat{x} \in T_x\mathbb{P}(V)$, and choose $v, \widehat{v} \in V$ such that

$$x := [v], \qquad \widehat{x} := [\widehat{v}], \qquad |v| = \sqrt{2\hbar}, \qquad \langle v, \widehat{v} \rangle = 0.$$

The infinitesimal action $L_x : \mathfrak{g} \to T_x\mathbb{P}(V)$ is given by $L_x\xi = [\xi v]$ for $\xi \in \mathfrak{g}$ and the symplectic form at x is

$$\omega_x(L_x\xi, \widehat{x}) = \langle \mathbf{i}\xi v, \widehat{v} \rangle.$$

Now choose a smooth path $\mathbb{R} \to V : t \mapsto v(t)$ such that

$$v(0) = v, \qquad \dot{v}(0) = \widehat{v}, \qquad |v(t)| = \sqrt{2\hbar}$$

for all t. Differentiate the curve $t \mapsto \langle \mu(x(t)), \xi \rangle$ at $t = 0$ to obtain

$$\begin{aligned} \langle d\mu(x)\widehat{x}, \xi \rangle &= \left.\frac{d}{dt}\right|_{t=0} \langle \mu(x(t)), \xi \rangle \\ &= \left.\frac{d}{dt}\right|_{t=0} \hbar \frac{\langle \mathbf{i}\xi v(t), v(t) \rangle}{|v(t)|^2} \\ &= \left.\frac{d}{dt}\right|_{t=0} \tfrac{1}{2} \langle \mathbf{i}\xi v(t), v(t) \rangle \\ &= \langle \mathbf{i}\xi v, \widehat{v} \rangle \\ &= \omega_x(L_x\xi, \widehat{x}). \end{aligned}$$

This proves Lemma 8.2. □

Lemma 8.3 *Let* $\mu : \mathbb{P}(V) \to \mathfrak{g}^*$ *be the moment map in Lemma 8.2. Then the Kempf–Ness function* $\Phi_x : \mathrm{G}^c \to \mathbb{R}$ *associated to* $x = [v]$ *is given by*

$$\Phi_x(g) = \hbar\Big(\log |g^{-1}v| - \log |v| \Big). \tag{8.2}$$

Proof Define $\Phi_x : \mathrm{G}^c \to \mathbb{R}$ by (8.2). Then $\Phi_x(u) = 0$ for all $u \in \mathrm{G}$ and

$$\begin{aligned} d\Phi_x(g)\widehat{g} &= \hbar \frac{\langle g^{-1}v, -g^{-1}\widehat{g}g^{-1}v\rangle}{|g^{-1}v|^2} \\ &= -\hbar \frac{\langle g^{-1}v, \mathbf{i}\mathrm{Im}(g^{-1}\widehat{g})g^{-1}v\rangle}{|g^{-1}v|^2} \\ &= -\langle \mu(g^{-1}x), \mathrm{Im}(g^{-1}\widehat{g})\rangle \end{aligned}$$

for $g \in \mathrm{G}^c$ and $\widehat{g} \in T_g\mathrm{G}^c$. Thus Φ_x is the Kempf–Ness function. □

Lemma 8.4 *Let $\mu : \mathbb{P}(V) \to \mathfrak{g}^*$ be the moment map in Lemma 8.2 and fix an element*

$$x = [v] \in \mathbb{P}(V).$$

Then the following holds.

(i) *Let $\zeta \in \mathscr{T}^c$, denote by*

$$\lambda_1 < \cdots < \lambda_k$$

the eigenvalues of $\mathbf{i}\zeta$ (understood as a linear operator on V), and denote by $V_i \subset V$ the corresponding eigenspaces. Write

$$v = \sum_{i=1}^{k} v_i, \qquad v_i \in V_i.$$

Then

$$w_\mu(x, \zeta) = \hbar \max_{v_i \neq 0} \lambda_i. \tag{8.3}$$

(ii) *Assume v is semistable, i.e. $0 \notin \overline{\mathrm{G}^c(v)}$, and choose an element $\zeta \in \mathfrak{g}^c$ such that v is an eigenvector of ζ. Then $\zeta v = 0$.*

Proof Write $\zeta = \xi + \mathbf{i}\eta \in \mathscr{T}^c$ with $\xi, \eta \in \mathfrak{g}$. Then

$$\begin{aligned} w_\mu(x, \zeta) &:= \lim_{t\to\infty} \langle \mu(\exp(\mathbf{i}t\zeta)x, \xi\rangle \\ &= \hbar \lim_{t\to\infty} \frac{\langle \exp(\mathbf{i}t\zeta)v, \mathbf{i}\xi \exp(\mathbf{i}t\zeta)v\rangle}{|\exp(\mathbf{i}t\zeta)v|^2} \\ &= \hbar \lim_{t\to\infty} \frac{\langle \exp(\mathbf{i}t\zeta)v, \mathbf{i}\zeta \exp(\mathbf{i}t\zeta)v\rangle}{|\exp(\mathbf{i}t\zeta)v|^2} \end{aligned}$$

$$= \hbar \lim_{t\to\infty} \frac{\langle \sum_{i=1}^k e^{\lambda_i t} v_i, \sum_{j=1}^k \lambda_j e^{\lambda_j t} v_j \rangle}{|\exp(\mathbf{i}t\zeta)v|^2}$$

$$= \hbar \lim_{t\to\infty} \frac{\sum_{i,j=1}^k \lambda_j e^{(\lambda_i+\lambda_j)t} \langle v_i, v_j \rangle}{\sum_{i,j=1}^k e^{(\lambda_i+\lambda_j)t} \langle v_i, v_j \rangle}$$

$$= \hbar \max_{v_i \neq 0} \lambda_i.$$

This proves (i). Now assume

$$0 \notin \overline{\mathrm{G}^c(v)}$$

and let $\zeta \in \mathfrak{g}^c$ and $\lambda \in \mathbb{C}$ such that

$$\zeta v = \lambda v.$$

Then the curves

$$e^{\lambda t} v = \exp(t\zeta)v, \qquad e^{\mathbf{i}\lambda t} v = \exp(\mathbf{i}t\zeta)v$$

in $\mathrm{G}^c(v)$ cannot contain 0 in their closure, and hence $\lambda = 0$. This proves (ii) and Lemma 8.4. □

The next result is the original Kempf–Ness theorem in [52].

Theorem 8.5 (Kempf–Ness) *Let* $\mu : \mathbb{P}(V) \to \mathfrak{g}^*$ *be the moment map in Lemma 8.2 and fix an element*

$$x = [v] \in \mathbb{P}(V).$$

Then the following holds.

(i) *v is unstable if and only if x is μ-unstable.*
(ii) *v is semistable if and only if x is μ-semistable.*
(iii) *v is polystable if and only if x is μ-polystable.*
(iv) *v is stable if and only if x is μ-stable.*

Proof The Kempf–Ness function $\Phi_x(\pi(g)) = \hbar(\log|g^{-1}v| - \log|v|)$ is unbounded below if and only if $0 \in \overline{\mathrm{G}^c(v)}$. By Theorem 7.3, Φ_x is unbounded below if and only if x is μ-unstable. This proves (i) and (ii).

To prove (iii) assume first that v is polystable. Thus $\mathrm{G}^c(v)$ is a closed subset of V and, in particular, $c := \inf_{g\in\mathrm{G}^c} |g^{-1}v| > 0$. Choose a sequence $g_i \in \mathrm{G}^c$ such that the sequence $|g_i^{-1}v|$ converges to c. Then the sequence $g_i^{-1}v \in V$ is bounded and hence has a convergent subsequence. Since $\mathrm{G}^c(v)$ is closed, the limit of this subsequence has the form $g^{-1}v$ for some $g \in \mathrm{G}^c$. Thus $|g^{-1}v| = c$ and so $\pi(g) \in \mathrm{G}^c/\mathrm{G}$ is a critical point of Φ_x. Hence x is μ-polystable by part (iii) of Theorem 7.3.

Conversely, assume x is μ-polystable. Choose a sequence $g_i \in \mathrm{G}^c$ such that the limit

$$w := \lim_{i\to\infty} g_i^{-1} v$$

exists. Since Φ_x has a critical point, there is a sequence $h_i \in \mathrm{G}^c_{x,0}$ such that $h_i g_i$ has a convergent subsequence (see part (viii) of Theorem 4.3). Pass to a subsequence so that the limit

$$g := \lim_{i\to\infty} h_i g_i$$

exists. Since $h_i^{-1} v = v$ for all i, by part (ii) of Lemma 8.4, we have

$$w = \lim_{i\to\infty} g_i^{-1} v = \lim_{i\to\infty} g_i^{-1} h_i^{-1} v = g^{-1} v \in \mathrm{G}^c(v).$$

Thus $\mathrm{G}^c(v)$ is closed and so v is polystable. This proves (iii).

Since the identity componenets of the isotropy subgroups of x and v agree in the semistable case, by part (ii) of Lemma 8.4, part (iv) follows from (iii). This proves Theorem 8.5. □

Remark 8.6 In Definition 8.1 the space $\mathbb{P}(V)$ can be replaced by any G^c-invariant closed complex submanifold $X \subset \mathbb{P}(V)$. Theorem 8.5 then continues to hold for the complex and symplectic structures and moment map on X obtained by restriction.

Chapter 9
Rationality

The definition of stability extends to actions of G^c by holomorphic automorphisms on a holomorphic line bundle E over a closed complex manifold X. In Definition 8.1 replace the complement of the origin in V by the complement of the zero section in E and replace the relation

$$x = [v]$$

by

$$v \in E_x \setminus \{0\}.$$

In the relevant applications the dual bundle $E^{-1} = \mathrm{Hom}(E, \mathbb{C})$ is ample.

Definition 9.1 A **linearization** of a holomorphic action of G^c on a complex manifold X consists of a holomorphic line bundle

$$E \to X$$

and a lift of the action of G^c on X to an action on E by holomorphic line bundle automorphisms.

A linearization of the G^c-action is required for the definition of stability in the intrinsic algebraic geometric setting. From the symplectic viewpoint the choice of the line bundle E is closely related to the symplectic form ω and the choice of the lift of the G^c-action on X to a G^c-action on E is closely related to the choice of the moment map μ. To obtain a linearization as in Definition 9.1 we will need to impose some additional conditions on the triple (X, ω, μ). Throughout we denote by

$$\mathbb{D} := \{z \in \mathbb{C} \mid |z| \leq 1\}$$

© The Author(s), under exclusive license to Springer Nature Switzerland AG 2021

V. Georgoulas et al., *The Moment-Weight Inequality and the Hilbert–Mumford Criterion*, Lecture Notes in Mathematics 2297,
https://doi.org/10.1007/978-3-030-89300-2_9

the closed unit disc in the complex plane and by

$$Z(\mathfrak{g}) := \{\xi \in \mathfrak{g} \mid [\xi, \eta] = 0 \; \forall \eta \in \mathfrak{g}\}$$

the center of $\mathfrak{g}$.

Let $x_0 \in X$ and let $u : \mathbb{R}/\mathbb{Z} \to \mathrm{G}$ be a smooth loop. Then the loop

$$u^{-1}x_0 : \mathbb{R}/\mathbb{Z} \to X$$

is contractible. (It is homotopic to a loop of the form $t \mapsto \exp(-t\xi)x_0$ with $\xi \in \Lambda$. This is a periodic orbit of a time independent periodic Hamiltonian system and the corresponding Hamiltonian function has a critical point. Connect x_0 to that critical point by a curve to obtain a homotopy to a constant loop.) If

$$\overline{x} : \mathbb{D} \to X$$

is a smooth map satisfying

$$\overline{x}(e^{2\pi \mathbf{i} t}) = u(t)^{-1}x_0, \tag{9.1}$$

define the **equivariant symplectic action** of the triple $(x_0, u, \overline{x})$ by

$$\mathscr{A}_\mu(x_0, u, \overline{x}) := -\int_{\mathbb{D}} \overline{x}^*\omega + \int_0^1 \langle \mu(u^{-1}x_0), u^{-1}\dot{u}\rangle \, dt. \tag{9.2}$$

The equivariant symplectic action depends only on the homotopy class of the loop $u : \mathbb{R}/\mathbb{Z} \to \mathrm{G}$ (see part (i) of Theorem 9.6 below).

Definition 9.2 Fix a constant $\hbar > 0$. The triple (X, ω, μ) is called **rational (with factor** $\hbar$**)** iff the following holds.

(A) The cohomology class of ω lifts to $H^2(X; 2\pi\hbar\mathbb{Z})$.
(B) $\mathscr{A}_\mu(x_0, u, \overline{x}) \in 2\pi\hbar\mathbb{Z}$ for all triples $(x_0, u, \overline{x})$ satisfying (9.1).

Theorem 9.3 *Let (X, ω, J) be a closed Kähler manifold equipped with a Hamiltonian group action by a compact Lie group* G *which is generated by an equivariant moment $\mu : X \to \mathfrak{g}$. Then the following holds.*

(i) *Condition (A) holds if and only if there exists a Hermitian line bundle $E \to X$ and a Hermitian connection ∇ on E such that the curvature of ∇ is*

$$F^\nabla = \frac{\mathbf{i}}{\hbar}\omega. \tag{9.3}$$

(ii) *Assume condition (A) and let (E, ∇) be as in (i). Assume further that* G *is connected. Then condition (B) holds if and only if there is a lift of the* G*-action on X to a* G*-action on E such that, for all $x_0 \in X$, $v_0 \in E_{x_0}$, and $u : \mathbb{R} \to \mathrm{G}$, the section*

$$v := u^{-1} v_0 \in E_x$$

along the path $x := u^{-1} x_0$ satisfies

$$\nabla_t v = \frac{\mathbf{i}}{\hbar} \langle \mu(x), u^{-1} \dot{u} \rangle v. \tag{9.4}$$

Proof See page 67. □

Remark 9.4 The complex line bundle $E \to X$ in Theorem 9.3 has the real first Chern class

$$c_1(E) = \left[-\frac{\omega}{2\pi\hbar} \right] \in H^2(X; \mathbb{R}).$$

Moreover, the curvature of ∇ is a $(1, 1)$-form and hence the Cauchy–Riemann operator

$$\bar{\partial}^{\nabla} : \Omega^{0,0}(X, E) \to \Omega^{0,1}(X, E)$$

defines a holomorphic structure on E. Conversely, for every holomorphic structure on E there exists a complex gauge transformation $g : X \to \mathbb{C}^*$ and a Hermitian connection ∇ on E such that

$$\bar{\partial}^{\nabla} = g \circ \bar{\partial} \circ g^{-1}, \qquad F^{\nabla} = \frac{\mathbf{i}}{\hbar} \omega.$$

In other words, the map $\nabla \mapsto \bar{\partial}^{\nabla}$ descends to an isomorphism from the moduli space of unitary gauge equivalence classes of Hermitian connections satisfying (9.3), which is isomorphic to the torus $H^1(X; \mathbf{i}\mathbb{R})/H^1(X; 2\pi\mathbf{i}\mathbb{Z})$, to the Jacobian of holomorphic structures on E. This isomorphism is analogous to the correspondence between symplectic and complex quotients in GIT.

Corollary 9.5 *Assume $\mathbb{P}(V)$ is equipped with a* G^c*-action, Fubini–Study form ω, and moment map μ as in Chap. 8 and let X be a* G*-invariant complex submanifold of $\mathbb{P}(V)$. Then (X, ω, μ) is rational.*

Proof Condition (A) holds as was noted in the text immediately after Definition 8.1. That condition (B) holds follows by inserting the formula (8.1) for the moment map in Eq. (9.2) and choosing $u : \mathbb{R}/\mathbb{Z} \to \mathrm{G} \subset \mathrm{U}(V)$ to be a loop of the form $u(t) = \exp(t\xi)$ with $\xi \in \Lambda$. □

Proof of Theorem 9.3 Part (i) follows directly from Chern–Weil theory and the classification of complex line bundles by their first Chern class. The proof of part (ii) has two steps.

Step 1 *Let $x_0 \in X$, $u : \mathbb{R}/\mathbb{Z} \to \mathrm{G}$, $\overline{x} : \mathbb{D} \to X$ such that*

$$\overline{x}(e^{2\pi \mathbf{i} t}) = u(t)^{-1} x_0.$$

Define $x(t) := u(t)^{-1} x_0$ and let $v(t) \in E_{x(t)}$ be a horizontal section. Then

$$v(1) = e^{-\frac{\mathbf{i}}{\hbar}\int_{\mathbb{D}} \overline{x}^*\omega} v(0).$$

Assume without loss of generality that $v_0 := v(0) \neq 0$, that $\overline{x}(s) = x_0$ for $0 \leq s \leq 1$, and that $\overline{x}$ is constant near the origin. Define $z(s,t) := \overline{x}(se^{2\pi \mathbf{i} t})$ for $0 \leq s, t \leq 1$ and let $v(s,t) \in E_{z(s,t)}$ be the solution of

$$\nabla_t v = 0, \qquad v(s,0) = v_0.$$

Since $v \neq 0$, there exists a unique function $\lambda : [0,1]^2 \to \mathbb{C}$ such that $\nabla_s v = \lambda v$. It satisfies $\lambda(s,0) = 0$ and

$$(\partial_t \lambda) v = \nabla_t(\lambda v) = \nabla_t \nabla_s v = \nabla_s \nabla_t v - F^{\nabla}(\partial_s z, \partial_t z) v = -\frac{\mathbf{i}}{\hbar}\omega(\partial_s z, \partial_t z) v.$$

Hence

$$\lambda(s,1) = -\frac{\mathbf{i}}{\hbar}\int_0^1 \omega(\partial_s z, \partial_t z)\, dt$$

and hence

$$\nabla_s v(s,1) = \left(-\frac{\mathbf{i}}{\hbar}\int_0^1 \omega(\partial_s z, \partial_t z)\, dt\right) v(s,1).$$

Thus $v(1,1) = e^{\mathbf{i}\alpha} v(0,1) = e^{\mathbf{i}\alpha} v_0$, where

$$\mathbf{i}\alpha = -\frac{\mathbf{i}}{\hbar}\int_0^1\int_0^1 \omega(\partial_s z, \partial_t z)\, dt\, ds = -\frac{\mathbf{i}}{\hbar}\int_{\mathbb{D}} \overline{x}^*\omega.$$

This proves Step 1.

Step 2 *Let $x_0 \in X$ and $u : \mathbb{R}/\mathbb{Z} \to \mathrm{G}$, and define the loop $x : \mathbb{R}/\mathbb{Z} \to X$ by*

$$x(t) := u(t)^{-1} x_0.$$

If $v(t) \in E_{x(t)}$ *is a section of* E *along* x *satisfying* (9.4) *then*

$$v(1) = e^{\frac{\mathbf{i}}{\hbar}\mathscr{A}_\mu(x_0,u,\overline{x})} v(0).$$

Let $v_0(t) \in E_{x(t)}$ be the horizontal lift with

$$v_0(0) = v(0) =: v_0.$$

Then $\nabla_t v_0 \equiv 0$ and there is a function $\lambda : [0, 1] \to \mathbb{C}$ such that $v = \lambda v_0$. It satisfies

$$\dot\lambda v_0 = \nabla_t(\lambda v_0) = \nabla_t v = \frac{\mathbf{i}}{\hbar}\langle \mu(x), u^{-1}\dot u\rangle v = \frac{\mathbf{i}}{\hbar}\langle \mu(x), u^{-1}\dot u\rangle \lambda v_0.$$

Hence

$$\lambda^{-1}\dot\lambda = \frac{\mathbf{i}}{\hbar}\langle \mu(x), u^{-1}\dot u\rangle, \qquad \lambda(0) = 1,$$

and hence

$$\lambda(1) = e^{\frac{\mathbf{i}}{\hbar}\int_0^1 \langle \mu(u^{-1}x_0), u^{-1}\dot u\rangle\, dt}.$$

By Step 1 and (B) this implies $v(1) = \lambda(1)v_0(1) = e^{\frac{\mathbf{i}}{\hbar}\mathscr{A}_\mu(x_0,u,\overline{x})}v(0)$. This proves Step 2.

Since G is connected it follows from Step 2 that the G-action on X lifts to a G-action on E via (9.4) if and only if $\mathscr{A}_\mu(x_0, u, \overline{x}) \in 2\pi\hbar\mathbb{Z}$ for all triples $(x_0, u, \overline{x})$. This proves Theorem 9.3. □

The following theorem shows that the equivariant symplectic action (9.2) is a homotopy invariant and gives conditions on the symplectic form ω under which the moment map μ can be chosen so that the triple (X, ω, μ) is rational.

Theorem 9.6 *Assume* G *is connected. Then the following holds.*

(i) *The action integral* (9.2) *is invariant under homotopy.*
(ii) *There is an* $N \in \mathbb{N}$ *such that* $N\alpha = 0$ *for each torsion class* $\alpha \in H_1(\mathrm{G}; \mathbb{Z})$.
(iii) *Let* $N \in \mathbb{N}$ *be as in (ii). If* $\langle \omega, \pi_2(X)\rangle \subset 2\pi\hbar N\mathbb{Z}$ *then there exists a central element* $\tau \in \mathrm{Z}(\mathfrak{g})$ *such that the moment map* $\mu + \tau$ *satisfies (B).*
(iv) *Assume* (X, ω, μ) *is rational with factor* $\hbar$ *and let* $\tau \in \mathrm{Z}(\mathfrak{g})$ *so* $\mu + \tau$ *is an equivariant moment map. Then* $(X, \omega, \mu + \tau)$ *is rational with factor* $\hbar$ *if and only if* $\langle \tau, \xi\rangle \in 2\pi\hbar\mathbb{Z}$ *for every* $\xi \in \Lambda$.

Proof Choose maps $x_0 : \mathbb{R} \to X$, $u : \mathbb{R} \times \mathbb{R}/\mathbb{Z} \to \mathrm{G}$, $\overline{x} : \mathbb{R} \times \mathbb{D} \to X$ such that

$$\overline{x}(s, e^{2\pi\mathbf{i}t}) = u(s, t)^{-1}x_0(s) =: x(s, t)$$

and define

$$\xi_s := u^{-1}\partial_s u, \qquad \xi_t := u^{-1}\partial_t u.$$

Then $\partial_s\xi_t - \partial_t\xi_s + [\xi_s, \xi_t] = 0$ and $\partial_t x = -L_x\xi_t$. Differentiate the function

$$a(s) := \mathscr{A}_\mu(x(s), u(s,\cdot), \overline{x}(s,\cdot)) = -\int_{\mathbb{D}} \overline{x}(s,\cdot)^*\omega + \int_0^1 \langle\mu(x), \xi_t\rangle\, dt.$$

Then

$$\begin{aligned}
\dot{a} &= \int_0^1 \Big(-\omega(\partial_s x, \partial_t x) + \langle d\mu(x)\partial_s x, \xi_t\rangle + \langle\mu(x), \partial_s\xi_t\rangle\Big)\, dt \\
&= \int_0^1 \Big(\omega(\partial_t x, \partial_s x) + \omega(L_x\xi_t, \partial_s x) + \langle\mu(x), \partial_s\xi_t\rangle\Big)\, dt \\
&= \int_0^1 \langle\mu(x), \partial_s\xi_t\rangle\, dt \\
&= \int_0^1 \langle\mu(x), \partial_t\xi_s - [\xi_s, \xi_t]\rangle\, dt \\
&= \int_0^1 \Big(\langle\mu(x), \partial_t\xi_s\rangle + \langle[\mu(x), \xi_t], \xi_s\rangle\Big)\, dt \\
&= \int_0^1 \Big(\langle\mu(x), \partial_t\xi_s\rangle - \langle d\mu(x)L_x\xi_t, \xi_s\rangle\Big)\, dt \\
&= \int_0^1 \Big(\langle\mu(x), \partial_t\xi_s\rangle + \langle d\mu(x)\partial_t x, \xi_s\rangle\Big)\, dt \\
&= \int_0^1 \partial_t\langle\mu(x), \xi_s\rangle\, dt = 0.
\end{aligned}$$

This proves (i).

Assertion (ii) follows from the fact that the fundamental group of G is abelian and finitely generated. We prove (iii). Let $H^1(\mathrm{G})$ denote the space of harmonic 1-forms on G with respect to the Riemannian metric induced by the invariant inner product on $\mathfrak{g}$. Then there is a vector space isomorphism $\mathrm{Z}(\mathfrak{g}) \to H^1(\mathrm{G}) : \tau \mapsto \alpha_\tau$ which assigns to each element $\tau \in \mathrm{Z}(\mathfrak{g})$ the harmonic 1-form $\alpha_\tau \in \Omega^1(\mathrm{G})$, given by $\alpha_\tau(u, \widehat{u}) := \langle\tau, u^{-1}\widehat{u}\rangle$. That the map $\tau \mapsto \alpha_\tau$ is surjective follows from the fact that the Ricci tensor of G is nonnegative, so every harmonic 1-form α is parallel.

Now let $\xi \in \Lambda$ and define the loop $u_\xi : \mathbb{R}/\mathbb{Z} \to \mathrm{G}$ by $u_\xi(t) := \exp(t\xi)$. An iterate of u_ξ is contractible if and only if

$$\int_{\mathbb{R}/\mathbb{Z}} u_\xi^*\alpha_\tau = \langle\tau, \xi\rangle = 0 \qquad \text{for all } \tau \in \mathrm{Z}(\mathfrak{g}),$$

or equivalently $\xi \in Z(\mathfrak{g})^\perp = [\mathfrak{g}, \mathfrak{g}]$. Thus the torsion classes in $\pi_1(G)$ correspond to $\Lambda \cap Z(\mathfrak{g})^\perp$ and the free part corresponds to $\Lambda \cap Z(\mathfrak{g})$.

Choose an integral basis $\xi_1, \dots, \xi_k \in Z(\mathfrak{g}) \cap \Lambda$ of $Z(\mathfrak{g})$ and a point $x_0 \in X$. Then each loop $\mathbb{R}/\mathbb{Z} \to X : t \mapsto \exp(-t\xi_j)x_0$ is contractible (see page 66). Hence there exist smooth maps $\overline{x}_j : \mathbb{D} \to X$ such that

$$\overline{x}_j(e^{2\pi \mathbf{i} t}) = \exp(-t\xi_j)x_0$$

for $t \in \mathbb{R}$ and $j = 1, \dots, k$. Define $\tau \in Z(\mathfrak{g})$ by

$$\langle \tau, \xi_j \rangle := \int_{\mathbb{D}} \overline{x}_j^* \omega - \langle \mu(x_0), \xi_j \rangle = -\mathscr{A}_\mu(x_0, u_{\xi_j}, \overline{x}_j), \qquad j = 1, \dots, k. \tag{9.5}$$

We claim that $\mu + \tau$ satisfies (B). By the definition of τ,

$$\mathscr{A}_{\mu+\tau}(x_0, u_{\xi_j}, \overline{x}_j) = 0, \qquad j = 1, \dots, k.$$

Since $\langle \omega, \pi_2(X) \rangle \subset 2\pi\hbar N\mathbb{Z}$ this implies that $\mathscr{A}_{\mu+\tau}(x_0, u_{\xi_j}, \overline{x}) \in 2\pi\hbar N\mathbb{Z}$ for every j and every smooth map $\overline{x} : \mathbb{D} \to X$ such that $\overline{x}(e^{2\pi \mathbf{i} t}) = \exp(-t\xi_j)x_0$. Now let $\xi \in \Lambda \cap Z(\mathfrak{g})^\perp$. Then the loop $\mathbb{R}/\mathbb{Z} \to X : t \mapsto \exp(-t\xi)x_0$ is contractible and hence there is a smooth map $\overline{x} : \mathbb{D} \to X$ such that

$$\overline{x}(e^{2\pi \mathbf{i} t}) = u_\xi(t)^{-1}x_0 = \exp(-t\xi)x_0$$

for all t. Define $\overline{x}_N(z) := \overline{x}(z^N)$. Since the action integral is invariant under homotopy, the loop $u_{N\xi}(t) = \exp(tN\xi)$ is contractible, and $\langle \omega, \pi_2(X) \rangle \subset 2\pi\hbar N\mathbb{Z}$, it follows that $\mathscr{A}_\mu(x_0, u_{N\xi}, \overline{x}_N) \in 2\pi\hbar N\mathbb{Z}$. Thus

$$\mathscr{A}_{\mu+\tau}(x_0, u_\xi, \overline{x}) = \mathscr{A}_\mu(x_0, u_\xi, \overline{x}) = \frac{1}{N}\mathscr{A}_\mu(x_0, u_{N\xi}, \overline{x}_N) \in 2\pi\hbar\mathbb{Z}.$$

Since the action integral is invariant under homotopy and additive under catenation it follows that $\mathscr{A}_{\mu+\tau}(x_0, u, \overline{x}) \in 2\pi\hbar\mathbb{Z}$ for every triple $(x_0, u, \overline{x})$. This proves (iii). Assertion (iv) follows directly from the definitions and this proves Theorem 9.6. □

Theorem 9.7 *Assume (X, ω, μ) is rational with factor $\hbar$ and G is connected. Let (E, ∇) be as in part (i) of Theorem 9.3. Then the following holds.*

(i) *The G-action on E in part (ii) of Theorem 9.3 extends uniquely to a G^c-action on E by holomorphic vector bundle automorphisms. Moreover,*

$$\nabla_t v = \frac{\mathbf{i}}{\hbar}\langle \mu(x), \xi \rangle v - \frac{1}{\hbar}\langle \mu(x), \eta \rangle v, \qquad \zeta := \xi + \mathbf{i}\eta := g^{-1}\dot{g}, \tag{9.6}$$

for all $x_0 \in X$, $v_0 \in E_{x_0}$, $g : \mathbb{R} \to G^c$ with $x := g^{-1}x_0$ and $v := g^{-1}v_0 \in E_x$.

(ii) *Let $x \in X$ and $v \in E_x \setminus \{0\}$. The Kempf–Ness function of x is*

$$\Phi_x(g) = \hbar(\log|g^{-1}v| - \log|v|).$$

(iii) *Let $x \in X$ and $\zeta \in \Lambda^c$ and define*

$$x^+ := \lim_{t\to\infty} \exp(\mathbf{i}t\zeta)x.$$

Then

$$w_\mu(x,\zeta) = \langle \mu(x^+), \mathrm{Re}(\zeta)\rangle \in 2\pi\hbar\mathbb{Z} \tag{9.7}$$

and the action of ζ on E_{x^+} is given by

$$\exp(-t\zeta)v^+ = e^{\frac{\mathbf{i}t}{\hbar}w_\mu(x,\zeta)}v^+. \tag{9.8}$$

Proof The Hermitian connection determines a G-invariant integrable complex structure on E via the splitting of the tangent bundle into horizontal and vertical subbundles. Hence the G-action on E extends to a G^c-action and this proves (i). Part (ii) follows by computing the derivative of the function

$$\Phi_x(g) := \hbar(\log|g^{-1}v| - \log|v|).$$

Equation (9.8) in (iii) follows from (9.6), and (9.7) follows from (9.8). This proves Theorem 9.7. □

Theorem 9.8 (Kempf–Ness) *Assume that (X, ω, μ) is rational with factor $\hbar$ and that G is connected. Let (E, ∇) be as in part (i) of Theorem 9.3 and let the G^c-action on E be as in Theorem 9.7. Let $\mathrm{Z} \subset E$ denote the zero section and let $x \in X$ and*

$$v \in E_x \setminus \{0\}.$$

Then

(i) *x is μ-unstable if and only if $\overline{\mathrm{G}^c(v)} \cap \mathrm{Z} \neq \emptyset$,*
(ii) *x is μ-semistable if and only if $\overline{\mathrm{G}^c(v)} \cap \mathrm{Z} = \emptyset$,*
(iii) *x is μ-polystable if and only if $\mathrm{G}^c(v)$ is a closed subset of E,*
(iv) *x is μ-stable if and only if $\mathrm{G}^c(v)$ is closed and G^c_v is discrete.*

Proof The proof is verbatim the same as that of Theorem 8.5. □

Definition 9.9 Fix a constant $\hbar > 0$. An invariant inner product on the Lie algebra $\mathfrak{g}$ is called **rational (with factor $\hbar$)** iff

$$\xi, \eta \in \Lambda, \quad [\xi, \eta] = 0 \quad \Longrightarrow \quad \langle \xi, \eta\rangle \in 2\pi\hbar\mathbb{Z}. \tag{9.9}$$

If G is a Lie subgroup of U(n) then an example of such an inner product is given by

$$\langle \xi, \eta \rangle := -2\pi\hbar \frac{\mathrm{trace}(\xi\eta)}{4\pi^2}$$

for $\xi, \eta \in \mathfrak{g} \subset \mathfrak{u}(n)$.

Theorem 9.10 *Assume that the triple (X, ω, μ) and the inner product on $\mathfrak{g}$ are rational with factor $\hbar$. Then, for every critical point $x \in X$ of the square of the moment map, there exists an integer k such that $k\mu(x) \in \Lambda$, i.e.*

$$L_x\mu(x) = 0 \qquad \Longrightarrow \qquad \mu(x) \in \mathbb{Q}\Lambda.$$

Proof First observe that

$$x \in X, \quad \xi \in \Lambda, \quad L_x\xi = 0 \qquad \Longrightarrow \qquad \langle \mu(x), \xi \rangle \in 2\pi\hbar\mathbb{Z}. \tag{9.10}$$

To see this fix an element $x \in X$ and an element $\xi \in \Lambda$ such that $L_x\xi = 0$. Then $\exp(\mathbf{i}t\xi)x = x$ for all t and hence $\langle \mu(x), \xi \rangle = w_\mu(x, \xi) \in 2\pi\hbar\mathbb{Z}$ by part (iii) of Theorem 9.7. This proves (9.10).

Now let $x \in X$ be a critical point of the square of the moment map, so

$$L_x\mu(x) = 0.$$

Let

$$\mathrm{T} := \overline{\{\exp(t\mu(x)) \,|\, t \in \mathbb{R}\}} \subset \mathrm{G}$$

be the torus generated by $\mu(x)$. Then the Lie algebra $\mathfrak{t} := \mathrm{Lie}(\mathrm{T})$ is contained in the kernel of L_x. Choose an integral basis $\xi_1, \ldots, \xi_k$ of $\mathfrak{t} \cap \Lambda$. Since $\mu(x) \in \mathfrak{t}$ there exist real numbers $\lambda_1, \ldots, \lambda_k$ such that

$$\mu(x) = \sum_{i=1}^{k} \lambda_i \xi_i.$$

Since $L_x\xi_j = 0$, it then follows from (9.10) that

$$\sum_{i=1}^{k} \lambda_i \frac{\langle \xi_i, \xi_j \rangle}{2\pi\hbar} = \frac{\langle \mu(x), \xi_j \rangle}{2\pi\hbar} \in \mathbb{Z}$$

for all j. Since the integer matrix $((2\pi\hbar)^{-1}\langle \xi_i, \xi_j \rangle)_{i,j=1}^{k}$ is invertible, it follows that $\lambda_i \in \mathbb{Q}$ for all i. Hence $\mu(x) \in \mathbb{Q}\Lambda$ and this proves Theorem 9.10. □

Chapter 10
The Dominant μ-Weight

The moment-weight inequality in Theorem 6.7 can be stated in the form

$$\sup_{0\neq\xi\in\mathfrak{g}} \frac{-w_\mu(x,\xi)}{|\xi|} \leq \inf_{g\in \mathrm{G}^c} |\mu(gx)|. \tag{10.1}$$

The main results of the present chapter assert that the supremum on the left in (10.1) is always attained (Theorem 10.1), that it is attained at a unique element ξ_0 up to scaling whenever x is μ-unstable (Theorem 10.2), that the inequality is strict in the μ-stable case (Theorem 10.3) and that equality holds in (10.1) in the μ-unstable case (Theorem 10.4). That equality also holds when x is μ-semistable, but not μ-stable, will follow from the Hilbert–Mumford criterion (Corollary 12.7). In [51] Kempf proved that the supremum in (10.1) is attained at a unique element $\xi_0 \in \Lambda$ up to scaling, whenever x is μ-unstable and the triple (X, ω, μ) and the inner product on $\mathfrak{g}$ are rational (Corollary 10.6). In general, the ray in $\mathfrak{g} \setminus \{0\}$ along which the supremum is attained need not intersect Λ.

Theorem 10.1 *Let $x_0 \in X$. Then there exists an element $\xi_0 \in \mathfrak{g}$ such that*

$$|\xi_0| = 1, \qquad -w_\mu(x_0,\xi_0) = \sup_{0\neq\xi\in\mathfrak{g}} \frac{-w_\mu(x_0,\xi)}{|\xi|}. \tag{10.2}$$

Proof See page 85. □

Theorem 10.2 (Generalized Kempf Uniqueness Theorem) *Assume that $x_0 \in X$ is μ-unstable. Then $\xi_0 \in \mathfrak{g}$ is uniquely determined by* (10.2).

Proof See page 85. □

© The Author(s), under exclusive license to Springer Nature Switzerland AG 2021

V. Georgoulas et al., *The Moment-Weight Inequality and the Hilbert–Mumford Criterion*, Lecture Notes in Mathematics 2297,
https://doi.org/10.1007/978-3-030-89300-2_10

Theorem 10.3 *Assume x_0 is μ-stable. Then*

$$0 < \lambda(x_0) := \inf_{0 \neq \xi \in \mathfrak{g}} \frac{w_\mu(x_0, \xi)}{|\xi|} \leq \inf_{x \in \overline{\mathrm{G}^c(x_0)} \setminus \mathrm{G}^c(x_0)} |\mu(x)|, \tag{10.3}$$

the set $\overline{\mathrm{G}^c(x_0)} \setminus \mathrm{G}^c(x_0)$ is compact, every $x \in \overline{\mathrm{G}^c(x_0)} \setminus \mathrm{G}^c(x_0)$ is μ-unstable, and

$$\inf_{g \in \mathrm{G}^c} |\mu(gx^+)| \geq \lambda(x_0)$$

for $\zeta \in \mathscr{T}^c$ and $x^+ := \lim_{t \to \infty} \exp(\mathbf{i}t\zeta)x_0$.

Proof See page 87. □

It follows from Theorem 5.3 and Lemma 5.7 that

$$\lambda(x_0) := \inf_{0 \neq \xi \in \mathfrak{g}} \frac{w_\mu(x_0, \xi)}{|\xi|} = \inf_{\zeta \in \mathscr{T}^c} \frac{w_\mu(x, \zeta)}{\sqrt{|\mathrm{Re}(\zeta)|^2 - |\mathrm{Im}(\zeta)|^2}} \tag{10.4}$$

for all $x_0 \in X$. Thus the function $\lambda : X \to \mathbb{R}$ is G^c-invariant. By Theorem 10.3 the number $\lambda(x_0)$ is positive whenever x_0 is μ-stable. In the work of Székelyhidi [72] the number $\lambda(x_0)$ is called the **modulus of stability** in the μ-stable case.

Another proof of Kempf Uniqueness for torus actions is contained in Lemma 11.2 in the next section. In the μ-unstable case an alternative proof of Theorem 10.1 is given in the following theorem, which also establishes equality in (10.1) and is the main result of this chapter.

Theorem 10.4 (Generalized Kempf Existence Theorem) *Assume that $x_0 \in X$ is μ-unstable so that*

$$m := \inf_{g \in \mathrm{G}^c} |\mu(gx_0)| > 0. \tag{10.5}$$

Let $x : \mathbb{R} \to X$ be the unique solution of (3.2) *and define $x_\infty := \lim_{t \to \infty} x(t)$. Let $g : \mathbb{R} \to \mathrm{G}^c$ be the solution of* (3.3) *so that $x(t) = g(t)^{-1}x_0$ for all $t \in \mathbb{R}$. Define the curves $\mathbb{R} \to \mathfrak{g} : t \mapsto \xi(t)$ and $\mathbb{R} \to \mathrm{G} : t \mapsto u(t)$ by*

$$g(t) =: \exp(-\mathbf{i}\xi(t))u(t). \tag{10.6}$$

Then the limit

$$\xi_\infty := \lim_{t \to \infty} \frac{\xi(t)}{t} \tag{10.7}$$

exists and satisfies

$$w_\mu(x_0, \xi_\infty) = -m^2, \qquad |\xi_\infty| = m. \tag{10.8}$$

Moreover, there exists an element $u_\infty \in \mathrm{G}$ *such that*

$$\xi_\infty = -u_\infty^{-1}\mu(x_\infty)u_\infty. \tag{10.9}$$

Proof See page 77. □

The proof of Theorem 10.4 is essentially due to Chen–Sun [23, Theorems 4.4 and 4.5]. They use the same argument to establish the existence of negative weights for linear actions on projective space under the assumption (10.5). Their χ is our ξ_∞ and their γ is the negative gradient flow line of the Kempf–Ness function going by the same name in the proof below.

Proof of Theorem 10.4 Let γ denote the image of g in $M = \mathrm{G}^c/\mathrm{G}$, i.e.

$$\gamma(t) := \pi(g(t)) = \pi(\exp(-\mathbf{i}\xi(t)))$$

for $t \in \mathbb{R}$. Let ∇ be the Levi-Civita connection on M. For $0 \le s < t$ let

$$\gamma_{s,t} : [s,t] \to M$$

be the geodesic connecting the points

$$\gamma_{s,t}(s) = \gamma(s), \qquad \gamma_{s,t}(t) = \gamma(t).$$

It is given by

$$\gamma_{s,t}(r) := \pi\left(g(s)\exp\left(-\mathbf{i}\frac{r-s}{t-s}\xi(s,t)\right)\right) \qquad \text{for } s \le r \le t, \tag{10.10}$$

where $\xi(s,t) \in \mathfrak{g}$ and $u(s,t) \in \mathrm{G}$ are chosen such that

$$g(s)\exp\big(-\mathbf{i}\xi(s,t)\big)u(s,t) = g(t). \tag{10.11}$$

(See Fig. 10.1.) For $0 \le s < t$ define the function $\rho_{s,t} : [s,t] \to [0,\infty)$ by

$$\rho_{s,t}(r) := d_M(\gamma_{s,t}(r), \gamma(r)) \qquad \text{for } s \le r \le t. \tag{10.12}$$

With this notation in place, we prove the assertions in nine steps.

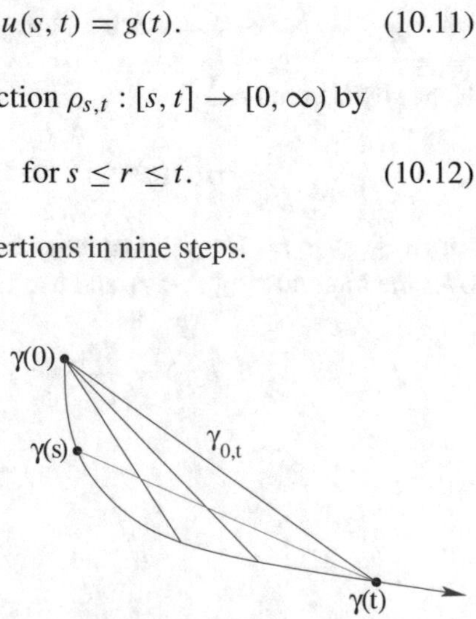

Fig. 10.1 A negative gradient flow line and geodesics

Step 1 *For every $t > 0$ we have*

$$|\nabla\dot{\gamma}(t)| = |L_{x(t)}{}^* L_{x(t)}\mu(x(t))|.$$

By (3.3), we have $g(t)^{-1}\dot{g}(t) = \mathbf{i}\mu(x(t)) \in \mathbf{i}\mathfrak{g}$ for all t. Hence Theorem C.1 asserts that

$$\nabla\dot{\gamma} = d\pi(g)g\mathbf{i}d\mu(x)\dot{x} = -d\pi(g)g\mathbf{i}L_x{}^* L_x\mu(x).$$

This proves Step 1.

Step 2 *There exist constants $c > 0$ and $0 < \varepsilon < 1$ such that, for all $t > 0$,*

$$\int_t^\infty |L_{x(r)}{}^* L_{x(r)}\mu(x(r))|\,dr \le \frac{c}{t^\varepsilon}.$$

This follows from Theorem 3.3 with $2/3 < \alpha < 1$ and $\varepsilon = (1-\alpha)/(2\alpha - 1)$.

Step 3 *Let c and ε be as in Step 2 and fix three real numbers r_0, s, t. Then*

$$0 \le s < r_0 < t, \quad \rho_{s,t}(r_0) \ne 0 \quad \Longrightarrow \quad \dot{\rho}_{s,t}(r_0) \le \frac{c}{r_0^\varepsilon}. \tag{10.13}$$

Assume $s < r_0 < t$ and $\rho_{s,t}(r_0) \ne 0$. Let r_1 be the smallest real number bigger than r_0 such that $\rho_{s,t}(r_1) = 0$. Thus

$$r_0 < r_1 \le t, \qquad \rho_{s,t}(r_1) = 0,$$

and

$$\rho_{s,t}(r) \ne 0 \qquad \text{for } r_0 \le r < r_1.$$

Hence by Lemma A.3

$$-\ddot{\rho}_{s,t}(r) \le |\nabla\dot{\gamma}(r)| = |L_{x(r)}{}^* L_{x(r)}\mu(x(r))|$$

for $r_0 \le r < r_1$. Here the equality follows from Step 1. Integrate this inequality over the interval $r_0 \le r < r_1$ and use Lemma A.3 to obtain

$$\begin{aligned}\dot{\rho}_{s,t}(r_0) &= \frac{d\rho_{s,t}}{dr^-}(r_1) - \int_{r_0}^{r_1} \ddot{\rho}_{s,t}(r)\,dr \\ &= -|\dot{\gamma}(r_1) - \dot{\gamma}_{s,t}(r_1)| - \int_{r_0}^{r_1} \ddot{\rho}_{s,t}(r)\,dr\end{aligned}$$

$$\leq \int_{r_0}^{r_1} |L_{x(r)}{}^* L_{x(r)}\mu(x(r))|\,dr$$
$$\leq \frac{c}{r_0^{\varepsilon}}.$$

Here the last inequality follows from Step 2. This proves Step 3.

Step 4 *Let $\varepsilon > 0$ be as in Step 2. Then there exists a constant $C > 0$ such that*

$$\rho_{s,t}(r) \leq C\left(r^{1-\varepsilon} - s^{1-\varepsilon}\right)$$

for all real numbers r, s, t such that $0 \leq s \leq r \leq t$.

Fix a real number r_1 such that $s < r_1 < t$. Suppose without loss of generality that $\rho_{s,t}(r_1) \neq 0$ and choose r_0 such that $s \leq r_0 < r_1$, $\rho_{s,t}(r_0) = 0$, and $\rho_{s,t}(r) \neq 0$ for $r_0 < r \leq r_1$. Now integrate the inequality

$$\dot{\rho}_{s,t}(r) \leq cr^{-\varepsilon}$$

in Step 3 over the interval $r_0 < r \leq r_1$ to obtain

$$\rho_{s,t}(r_1) = \int_{r_0}^{r_1} \dot{\rho}_{s,t}(r)\,dr \leq \int_{s}^{r_1} \frac{c\,dr}{r^{\varepsilon}} = \frac{c}{1-\varepsilon}\left({r_1}^{1-\varepsilon} - s^{1-\varepsilon}\right).$$

This proves Step 4 with $C := c/(1-\varepsilon)$.

Step 5 *Let $s \geq 0$. Then*

$$\left|\frac{\xi(s,t')}{t'-s} - \frac{\xi(s,t)}{t-s}\right| \leq \frac{C}{t^{\varepsilon}} \qquad \text{for all } t' \geq t \geq s+1$$

Thus the limit

$$\xi_\infty(s) := \lim_{t\to\infty} \frac{\xi(s,t)}{t-s}$$

exists in $\mathfrak{g}$.

The geodesics $\gamma_{s,t}$ and $\gamma_{s,t'}$ intersect at

$$\gamma(s) = \gamma_{s,t}(s) = \gamma_{s,t'}(s)$$

(see Fig. 10.1). Hence it follows from Eq. (10.10), Lemma A.4, and Step 4 that

$$\begin{aligned}\left|\frac{\xi(s,t')}{t'-s} - \frac{\xi(s,t)}{t-s}\right| &= |\dot{\gamma}_{s,t}(s) - \dot{\gamma}_{s,t'}(s)| \\ &\leq \frac{d_M\left(\gamma_{s,t}(t), \gamma_{s,t'}(t)\right)}{t-s}\end{aligned}$$

$$
\begin{aligned}
&= \frac{\rho_{s,t'}(t)}{t-s} \\
&\le C\frac{t^{1-\varepsilon}-s^{1-\varepsilon}}{t-s} \\
&\le \frac{C}{t^{\varepsilon}}
\end{aligned}
$$

for $t' \ge t \ge s+1$. This proves Step 5.

Step 6 $w_\mu(x(s), \xi_\infty(s)) = -m^2$ *for all* $s \ge 0$.
By Theorem 6.4, we have

$$|\mu(x_\infty)| = m.$$

Fix a real number $s \ge 0$ and define the geodesic

$$\gamma_{s,\infty} : [s,\infty) \to M$$

by

$$\gamma_{s,\infty}(r) := \pi(g(s)\exp(-\mathbf{i}(r-s)\xi_\infty(s))) = \lim_{t\to\infty} \gamma_{s,t}(r)$$

for $r \ge s$. By Step 4, we have

$$d_M\big(\gamma(r), \gamma_{s,t}(r)\big) = \rho_{s,t}(r) \le C\big(r^{1-\varepsilon} - s^{1-\varepsilon}\big)$$

for $s \le r \le t$. Take the limit $t \to \infty$ to obtain

$$d_M\big(\gamma(r), \gamma_{s,\infty}(r)\big) \le C\big(r^{1-\varepsilon} - s^{1-\varepsilon}\big) \tag{10.14}$$

for $r \ge s \ge 0$. Now the Kempf–Ness function is globally Lipschitz continuous with Lipschitz constant $L := \sup_{g\in G^c} |\mu(gx_0)|$. Hence it follows from (10.14) that

$$|\Phi_{x_0}(\gamma(t)) - \Phi_{x_0}(\gamma_{s,\infty}(t))| \le LC\big(t^{1-\varepsilon} - s^{1-\varepsilon}\big) \qquad \text{for } t \ge s \ge 0. \tag{10.15}$$

Integrate the equation

$$\frac{d}{dr}\Phi_{x_0}(\gamma(r)) = -|\mu(x(r))|^2$$

to obtain

$$\Phi_{x_0}(\gamma(t)) = \Phi_{x_0}(\gamma(s)) - \int_s^t |\mu(x(r))|^2\,dr. \tag{10.16}$$

By Lemma 5.2 and (10.15) and (10.16) we have

$$\begin{aligned}
w_\mu(x(s), \xi_\infty(s)) &= \lim_{t\to\infty} \frac{\Phi_{x_0}(\gamma_{s,\infty}(s+t))}{t} \\
&= \lim_{t\to\infty} \frac{\Phi_{x_0}(\gamma_{s,\infty}(t)) - \Phi_{x_0}(\gamma(s))}{t-s} \\
&= \lim_{t\to\infty} \frac{\Phi_{x_0}(\gamma(t)) - \Phi_{x_0}(\gamma(s))}{t-s} \\
&= -\lim_{t\to\infty} \frac{1}{t-s}\int_s^t |\mu(x(r))|^2\, dr \\
&= -|\mu(x_\infty)|^2 \\
&= -m^2.
\end{aligned}$$

This proves Step 6.

Step 7 $|\xi_\infty(s)| = m$ *for all* $s \geq 0$.

By definition of $\xi(s,t)$ in (10.11) we have

$$\begin{aligned}
\frac{|\xi(s,t)|}{t-s} &= \frac{d_M(\gamma(s),\gamma(t))}{t-s} \\
&\leq \frac{1}{t-s}\int_s^t |\dot\gamma(r)|\, dr \\
&= \frac{1}{t-s}\int_s^t |\mu(x(r))|\, dr.
\end{aligned}$$

Take the limit $t \to \infty$. Then

$$|\xi_\infty(s)| \leq \lim_{t\to\infty} \frac{1}{t-s}\int_s^t |\mu(x(r))|\, dr = |\mu(x_\infty)| = m.$$

Moreover, it follows from the moment-weight inequality in Theorem 6.7 that

$$m^2 = -w(x(s), \xi_\infty(s)) \leq |\xi_\infty(s)| \inf_{g\in \mathrm{G}^c} |\mu(gx_0)| = m|\xi_\infty(s)|$$

and hence $|\xi_\infty(s)| \geq m$. This proves Step 7.

Step 8 *For every* $s \geq 0$ *there exists an element* $u_\infty(s) \in \mathrm{G}$ *such that*

$$\xi_\infty(0) = u_\infty(s)^{-1}\xi_\infty(s)u_\infty(s).$$

The geodesics $\gamma_{s,t}$ and $\gamma_{0,t}$ intersect at the point $\gamma(t) = \gamma_{s,t}(t) = \gamma_{0,t}(t)$ (see Fig. 10.1). Hence it follows from (10.10), (10.11) and Lemma A.4 that

$$\begin{aligned}\left|u(s,t)^{-1}\frac{\xi(s,t)}{t-s}u(s,t) - u(0,t)^{-1}\frac{\xi(0,t)}{t}u(0,t)\right| &= |\dot{\gamma}_{s,t}(t) - \dot{\gamma}_{0,t}(t)| \\ &\le \frac{d_M(\gamma_{s,t}(s), \gamma_{0,t}(s))}{t-s} \\ &= \frac{\rho_{0,t}(s)}{t-s} \\ &\le \frac{Cs^{1-\varepsilon}}{t-s}\end{aligned}$$

for $t \ge s+1 \ge 0$. Here the last inequality follows from Step 4. Now choose a sequence $t_i \to \infty$ such that the limit $u_\infty(s) := \lim_{i\to\infty} u(s,t_i)u(0,t_i)^{-1}$ exists. Then, by Step 5, we have $\xi_\infty(0) = \lim_{i\to\infty} t_i^{-1}\xi(0,t_i)$ and hence

$$\begin{aligned}\xi_\infty(0) &= \lim_{i\to\infty} u(0,t_i)u(s,t_i)^{-1}\frac{\xi(s,t_i)}{t_i-s}u(s,t_i)u(0,t_i)^{-1} \\ &= u_\infty(s)^{-1}\xi_\infty(s)u_\infty(s).\end{aligned}$$

This proves Step 8.

Step 9 $\lim_{s\to\infty}\xi_\infty(s) = -\mu(x_\infty)$.

By Step 5 the inequality

$$\left|\frac{\xi(s,t')}{t'-s} - \frac{\xi(s,t)}{t-s}\right| \le \frac{C}{t^\varepsilon} \le \frac{C}{s^\varepsilon}$$

holds for $t' \ge t \ge s+1$. Take the limit $t' \to \infty$ to obtain

$$|\xi_\infty(s) - \xi(s,s+1)| \le \frac{C}{s^\varepsilon} \qquad \text{for all } s \ge 0.$$

Now suppose $\xi(s,s+1) + \mu(x(s)) \ne 0$. Then $\dot{\gamma}(s) \ne \dot{\gamma}_{s,s+1}(s)$, so there is a real number t such that $s < t \le s+1$, $\rho_{s,s+1}(t) = 0$, and $\rho_{s,s+1}(r) \ne 0$ for $s < r < t$. By Step 3 this implies $\dot{\rho}_{s,s+1}(r) \le cr^{-\varepsilon}$ for $s < r < t$ and hence, by Lemma A.3,

$$|\mu(x(s)) + \xi(s,s+1)| = |\dot{\gamma}(s) - \dot{\gamma}_{s,s+1}(s)| = \lim_{r\searrow s}\dot{\rho}_{s,s+1}(r) \le \frac{c}{s^\varepsilon}.$$

Hence

$$|\xi_\infty(s) + \mu(x(s))| \le \frac{c+C}{s^\varepsilon}$$

for all $s \ge 0$ and this proves Step 9.

The existence of the limit in (10.7) follows from Step 5, that it satisfies (10.8) follows from Steps 6 and 7, and that it satisfies (10.9) follows from Steps 8 and 9. This proves Theorem 10.4.

□

In preparation for the proofs of the remaining theorems in this chapter we establish a convergence result for the Kempf–Ness function.

Lemma 10.5 *Let $x_0 \in X$, define*

$$\lambda(x_0) := \inf_{0 \neq \xi \in \mathfrak{g}} \frac{w_\mu(x_0, \xi)}{|\xi|}, \tag{10.17}$$

denote by $\Phi_{x_0} : \mathrm{G}^c \to \mathbb{R}$ the lifted Kempf–Ness function, and define

$$S_t := \left\{ \exp(-\mathbf{i}t\xi) \,\middle|\, \xi \in \mathfrak{g},\ |\xi| = 1 \right\}$$

for $t > 0$. Then $\lambda(x_0) > -\infty$ and the following holds.

(i) $\frac{1}{t} \inf_{S_t} \Phi_{x_0} \leq \lambda(x_0)$ *for all $t > 0$.*

(ii) *There exists a sequence $t_i > 0$ of positive real numbers and a convergent sequence $\xi_i \in \mathfrak{g}$ such that*

$$|\xi_i| = 1, \qquad \Phi_{x_0}(\exp(-\mathbf{i}t_i\xi_i)) = \inf_{S_{t_i}} \Phi_{x_0}, \qquad \lim_{i\to\infty} t_i = \infty.$$

The limit $\xi_0 := \lim_{i\to\infty} \xi_i$ of any such sequence satisfies (10.2).

(iii) $\lim_{t\to\infty} t^{-1} \inf_{S_t} \Phi_{x_0} = \lambda(x_0)$.

Proof We prove part (i). The number $\lambda(x_0)$ in (10.17) is finite by the moment-weight inequality (6.13) in Theorem 6.7. Fix two real numbers $t > 0$ and $\varepsilon > 0$, and choose $\xi \in \mathfrak{g}$ such that

$$|\xi| = 1, \qquad w_\mu(x_0, \xi) < \lambda(x_0) + \varepsilon.$$

Since the function $s \mapsto \langle \mu(\exp(\mathbf{i}s\xi)x_0), \xi \rangle$ is nondecreasing and converges to the weight $w_\mu(x_0, \xi)$ as s tends to infinity, we have

$$\begin{aligned}
\frac{1}{t}\Phi_{x_0}(\exp(-\mathbf{i}t\xi)) &= \frac{1}{t}\int_0^t \langle \mu(\exp(\mathbf{i}s\xi)x_0), \xi \rangle \, ds \\
&\leq w_\mu(x_0, \xi) \\
&< \lambda(x_0) + \varepsilon.
\end{aligned}$$

Here the first equality follows from (4.14) and the fact that $\Phi_{x_0}(\mathbb{1}) = 0$. Since $\varepsilon > 0$ was chosen arbitrary, this proves (i).

We prove part (ii). Choose any sequence $t_i > 0$ that tends to infinity. Since Φ_{x_0} is continuous and S_{t_i} is compact, the axiom of countable choice asserts that there exists a sequence $\xi_i \in \mathfrak{g}$ such that

$$|\xi_i| = 1, \qquad \Phi_{x_0}(\exp(-\mathbf{i}t_i\xi_i)) = \inf_{S_{t_i}} \Phi_{x_0}$$

for all i. Passing to a subsequence, we may assume that the sequence $(\xi_i)_{i\in\mathbb{N}}$ converges. Denote the limit by $\xi_0 := \lim_{i\to\infty} \xi_i$. Then

$$|\xi_0| = \lim_{i\to\infty} |\xi_i| = 1.$$

Moreover, by Lemma 5.2, the function $t \mapsto t^{-1}\Phi_{x_0}(\exp(-\mathbf{i}t\xi_i))$ is nondecreasing and hence by part (i) we have

$$\frac{1}{t}\Phi_{x_0}(\exp(-\mathbf{i}t\xi_i)) \le \frac{1}{t_i}\inf_{S_{t_i}} \Phi_{x_0} \le \lambda(x_0) \qquad \text{for } 0 < t \le t_i.$$

Take the limit $i \to \infty$ to obtain

$$\frac{1}{t}\Phi_{x_0}(\exp(-\mathbf{i}t\xi_0)) \le \lambda(x_0)$$

for all $t > 0$. By Lemma 5.2, this implies

$$w_\mu(x_0, \xi_0) = \lim_{t\to\infty} \frac{1}{t}\Phi_{x_0}(\exp(-\mathbf{i}t\xi_0)) \le \lambda(x_0)$$

and hence

$$w_\mu(x_0, \xi_0) = \lambda(x_0)$$

by definition of $\lambda(x_0)$ in (10.17). This proves (ii).

We prove part (iii). Choose $t_i > 0$, $\xi_i \in \mathfrak{g}$, and $\xi_0 = \lim_{i\to\infty} \xi_i$ as in (ii). Then

$$\lambda(x_0) = w_\mu(x_0, \xi_0) = \lim_{t\to\infty} \frac{1}{t}\Phi_{x_0}(\exp(-\mathbf{i}t\xi_0))$$

by Lemma 5.2. Fix a constant $\varepsilon > 0$ and choose $t_0 > 0$ such that

$$\frac{1}{t_0}\Phi_{x_0}(\exp(-\mathbf{i}t_0\xi_0)) > \lambda(x_0) - \varepsilon.$$

Then there exists a constant $i_0 > 0$ such that, for all $i \in \mathbb{N}$,

$$i \ge i_0 \qquad \Longrightarrow \qquad t_i \ge t_0 \quad \text{and} \quad \frac{1}{t_0}\Phi_{x_0}(\exp(-\mathbf{i}t_0\xi_i)) > \lambda(x_0) - \varepsilon.$$

Hence, for every $i \in \mathbb{N}$ with $i \geq i_0$, we have

$$\begin{aligned}\lambda(x_0) - \varepsilon &< \frac{1}{t_0}\Phi_{x_0}(\exp(-\mathbf{i}t_0\xi_i)) \\ &\leq \frac{1}{t_i}\Phi_{x_0}(\exp(-\mathbf{i}t_i\xi_i)) \\ &= \frac{1}{t_i}\inf_{S_{t_i}}\Phi_{x_0} \\ &\leq \lambda(x_0).\end{aligned}$$

Here the second inequality holds because $t \mapsto t^{-1}\Phi_{x_0}(\exp(-\mathbf{i}t\xi_i))$ is nondecreasing by Lemma 5.2, the equality follows from the choice of the sequence ξ_i in part (ii), and the last inequality follows from part (i). Thus

$$\lim_{i\to\infty}\frac{1}{t_i}\inf_{S_{t_i}}\Phi_{x_0} = \lambda(x_0). \tag{10.18}$$

Now Lemma 5.2 asserts that

$$\frac{1}{t_i}\Phi_{x_0}(\exp(-\mathbf{i}t_i\xi)) \leq \frac{1}{t}\Phi_{x_0}(\exp(-\mathbf{i}t\xi))$$

for all $i \in \mathbb{N}$, all $t \geq t_i$, and all $\xi \in \mathfrak{g}$ with $|\xi| = 1$. Take the infimum over all $\xi \in \mathfrak{g}$ with $|\xi| = 1$ and use part (i) to obtain

$$\frac{1}{t_i}\inf_{S_{t_i}}\Phi_{x_0} \leq \frac{1}{t}\inf_{S_t}\Phi_{x_0} \leq \lambda(x_0) \qquad \text{for all } i \in \mathbb{N} \text{ and all } t \geq t_i.$$

By (10.18) this proves (iii) and Lemma 10.5. □

Proof of Theorem 10.1 This follows directly from part (ii) of Lemma 10.5. □

Proof of Theorem 10.2 Assume x_0 is μ-unstable. Then

$$\lambda(x_0) = \inf_{0\neq\xi\in\mathfrak{g}}\frac{w_\mu(x_0,\xi)}{|\xi|} \leq \frac{w_\mu(x_0,\xi_\infty)}{|\xi_\infty|} < 0$$

by Theorem 10.4. Define $m := -\lambda(x_0)$, let $\xi_0, \xi_1 \in \mathfrak{g}$ such that

$$|\xi_0| = |\xi_1| = 1, \qquad w_\mu(x_0,\xi_0) = w_\mu(x_0,\xi_1) = \lambda(x_0) = -m, \tag{10.19}$$

and consider the geodesics

$$\gamma_0(t) := \pi\big(\exp(-\mathbf{i}t\xi_0)\big), \qquad \gamma_1(t) := \pi\big(\exp(-\mathbf{i}t\xi_1)\big).$$

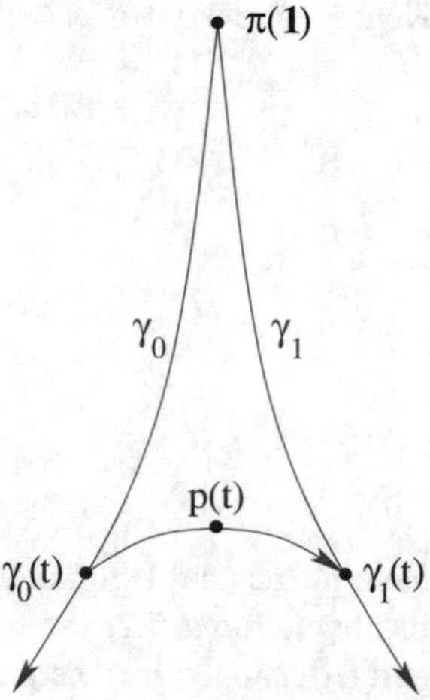

Fig. 10.2 Proof of Kempf uniqueness

(See Fig. 10.2.) For $t > 0$ choose $\eta(t) \in \mathfrak{g}$ and $u(t) \in \mathrm{G}$ such that

$$\exp(-\mathbf{i}t\xi_0)\exp(\mathbf{i}\eta(t)) = \exp(-\mathbf{i}t\xi_1)u(t)$$

and define

$$p(t) := \pi\big(\exp(-\mathbf{i}t\xi_0)\exp(\mathbf{i}\tfrac{1}{2}\eta(t))\big) \in M = \mathrm{G}^c/\mathrm{G}.$$

Then $p(t)$ is the midpoint of the geodesic joining $\gamma_0(t)$ and $\gamma_1(t)$. Hence it follows from Lemma A.6 that

$$\begin{aligned} d(\pi(\mathbb{1}), p(t))^2 &\leq \frac{d(\pi(\mathbb{1}), \gamma_0(t))^2 + d(\pi(\mathbb{1}), \gamma_1(t))^2}{2} - \frac{d(\gamma_0(t), \gamma_1(t))^2}{4} \\ &= t^2 - \frac{d(\gamma_0(t), \gamma_1(t))^2}{4} \\ &\leq t^2\left(1 - \frac{|\xi_0 - \xi_1|^2}{4}\right). \end{aligned}$$

The last inequality holds for $t \geq 1$, by Lemma A.4. Thus

$$\frac{r(t)}{t} \leq \sqrt{1 - \frac{|\xi_0 - \xi_1|^2}{4}}, \qquad r(t) := d(\pi(\mathbb{1}), p(t)), \tag{10.20}$$

for $t \geq 1$. Moreover,

$$\Phi_{x_0}(\gamma_0(t)) \leq -tm, \qquad \Phi_{x_0}(\gamma_1(t)) \leq -tm$$

for all $t \geq 0$ by Eq. (10.19) and Lemma 5.2. Since Φ_{x_0} is convex along geodesics, this implies

$$\Phi_{x_0}(p(t)) \leq -tm \tag{10.21}$$

for all $t \geq 0$. In particular, the function $r(t) = d(\pi(\mathbb{1}), p(t))$ diverges to infinity as t tends to infinity. Hence

$$\lim_{t\to\infty} \frac{1}{r(t)} \inf_{S_{r(t)}} \Phi_{x_0} = -m$$

by part (iii) of Lemma 10.5 and, for $t \geq 1$, we have

$$\frac{m}{\sqrt{1 - \frac{|\xi_0 - \xi_1|^2}{4}}} \leq \frac{tm}{r(t)} \leq -\frac{\Phi_{x_0}(p(t))}{r(t)} \leq -\frac{1}{r(t)} \inf_{S_{r(t)}} \Phi_{x_0}.$$

Here the first inequality follows from (10.20), the second from (10.21), and the last from the fact that $p(t) \in S_{r(t)}$. Take the limit $t \to \infty$ to obtain

$$\frac{m}{\sqrt{1 - \frac{|\xi_0 - \xi_1|^2}{4}}} \leq m$$

and hence $\xi_0 = \xi_1$. This proves Theorem 10.2. □

Proof of Theorem 10.3 Assume x_0 is μ-stable and define the real number $\lambda(x_0)$ by (10.17). Then $\lambda(x_0) > 0$ by Theorems 10.1 and 7.4. Now let

$$x \in \overline{\mathrm{G}^c(x_0)} \setminus \mathrm{G}^c(x_0).$$

Then there exist sequences $\xi_i \in \mathfrak{g}$, $t_i > 0$, and $u_i \in \mathrm{G}$ such that $t_i \to \infty$, $|\xi_i| = 1$ for all i, and $x = \lim_{i\to\infty} u_i \exp(\mathbf{i}t_i\xi_i)x_0$. Hence

$$\begin{aligned}
\frac{1}{t_i} \inf_{S_{t_i}} \Phi_{x_0} \leq \frac{1}{t_i} \Phi_{x_0}(\exp(-\mathbf{i}t_i\xi_i)) &= \frac{1}{t_i} \int_0^{t_i} \langle \mu(\exp(\mathbf{i}t\xi_i)x_0), \xi_i \rangle \, dt \\
&\leq \langle \mu(\exp(\mathbf{i}t_i\xi_i)x_0), \xi_i \rangle \\
&\leq |\mu(u_i \exp(\mathbf{i}t_i\xi_i)x_0)|
\end{aligned}$$

for all $i \in \mathbb{N}$. Take the limit $i \to \infty$ and use part (iii) of Lemma 10.5 to obtain the inequality $\lambda(x_0) \leq |\mu(x)|$. This proves (10.3).

It follows from (10.3) that every element $x \in \overline{\mathrm{G}^c(x_0)} \setminus \mathrm{G}^c(x_0)$ satisfies

$$\inf_{g\in\mathrm{G}^c} |\mu(gx)| \geq \lambda(x_0) > 0$$

and hence is μ-unstable. Since the set $X^{\mathrm{us}} \subset X$ of μ-unstable points is closed by Theorem 7.2, so is the set

$$\overline{\mathrm{G}^c(x_0)} \setminus \mathrm{G}^c(x_0) = \overline{\mathrm{G}^c(x_0)} \cap X^{\mathrm{us}},$$

and hence this set is compact. Now let $\zeta \in \mathscr{T}^c$. Then the limit point

$$x^+ := \lim_{t\to\infty} \exp(\mathbf{i}t\zeta)x_0$$

satisfies $L^c_{x^+}\zeta = 0$ by Lemma 5.4, and hence cannot belong to $\mathrm{G}^c(x_0)$. Thus

$$x^+ \in \overline{\mathrm{G}^c(x_0)} \setminus \mathrm{G}^c(x_0)$$

and so $\inf_{g\in\mathrm{G}^c} |\mu(gx^+)| \geq \lambda(x_0)$. This proves Theorem 10.3. □

Corollary 10.6 *Let $x_0 \in X$ be μ-unstable. Then there exists a unique element $\xi_0 \in \mathfrak{g}$ such that $|\xi_0| = 1$ and*

$$-\, w_\mu(x_0, \xi_0) = \inf_{g\in\mathrm{G}^c} |\mu(gx_0)| = \sup_{0\neq\xi\in\mathfrak{g}} \frac{-w_\mu(x_0, \xi)}{|\xi|}. \tag{10.22}$$

If the triple (X, ω, μ) and the inner product on $\mathfrak{g}$ are rational with factor $\hbar$ then there exists a positive integer ℓ such that $\sqrt{2\pi\hbar\ell}\xi_0 \in \Lambda$.

Proof Let ξ_∞ be as in Theorem 10.4. Then the element

$$\xi_0 := \frac{\xi_\infty}{|\xi_\infty|} \in \mathfrak{g}$$

satisfies the first equation in (10.22). The second equation in (10.22) follows from the first ($\leq$) and the moment-weight inequality in Theorem 6.7 ($\geq$). Uniqueness follows from Theorem 10.2. To prove the last assertion, let $k \in \mathbb{N}$ such that

$$k\mu(x_\infty) \in \Lambda$$

(Theorem 9.10) and take

$$\ell := \frac{k^2|\mu(x_\infty)|^2}{2\pi\hbar}.$$

Then $\ell \in \mathbb{N}$ and

$$\sqrt{2\pi\hbar\ell}\xi_0 = k|\mu(x_\infty)|\xi_0 = k\xi_\infty = -u_\infty^{-1}k\mu(x_\infty)u_\infty \in \Lambda.$$

This proves Corollary 10.6. □

Chapter 11
Torus Actions

Throughout this chapter we assume that the Lie group $\mathrm{G} = \mathrm{T}$ is a torus with Lie algebra $\mathfrak{t} := \mathrm{Lie}(\mathrm{T})$ and complexification T^c. The first main result is Theorem 11.1, which asserts that in the case of a torus action the function

$$\mathfrak{t} \setminus \{0\} \to \mathbb{R} : \xi \mapsto w_\mu(x, \xi) \tag{11.1}$$

is continuous for every $x \in X$. The second main result (Theorem 11.3) is not used elsewhere in this book. It asserts that the closure of the image of a complexified group orbit under the moment map is convex in the case of a torus action.

Theorem 11.1 (Continuity) *Let $x \in X$. Then*

$$w_\mu(x, \xi) = \sup_{g \in \mathrm{T}^c} \langle \mu(gx), \xi \rangle \tag{11.2}$$

for every $\xi \in \mathfrak{t} \setminus \{0\}$ and the function (11.1) *is continuous.*

Proof Let $x_0 \in X$ and $\xi \in \mathfrak{t} \setminus \{0\}$. Then

$$w_\mu(x_0, \xi) = \lim_{t \to \infty} \langle \mu(\exp(\mathbf{i}t\xi)x_0), \xi \rangle \le \sup_{g \in \mathrm{T}^c} \langle \mu(gx_0), \xi \rangle.$$

Now let $g \in \mathrm{T}^c$. Then the function $t \mapsto \langle \mu(\exp(\mathbf{i}t\xi)gx_0), \xi \rangle$ is nondecreasing by (2.7) and converges to $w_\mu(gx_0, \xi)$ as t tends to infinity. This implies

$$\langle \mu(gx_0), \xi \rangle \le w_\mu(gx_0, \xi) = w_\mu(gx_0, g\xi g^{-1}) = w_\mu(x_0, \xi)$$

© The Author(s), under exclusive license to Springer Nature Switzerland AG 2021
V. Georgoulas et al., *The Moment-Weight Inequality and the Hilbert–Mumford Criterion*, Lecture Notes in Mathematics 2297,
https://doi.org/10.1007/978-3-030-89300-2_11

by Theorem 5.3. Hence

$$\sup_{g\in \mathrm{T}^c} \langle \mu(gx_0), \xi\rangle \le w_\mu(x_0, \xi)$$

and this proves (11.2) for $x = x_0$.

Now assume, by contradiction, that the function (11.1) with $x = x_0$ is not continuous. Then there exists a sequence $\xi_i \in \mathfrak{t} \setminus \{0\}$ converging to an element $\xi \in \mathfrak{t} \setminus \{0\}$ such that the sequence $w_\mu(x_0, \xi_i)$ does not converge to $w_\mu(x_0, \xi)$. Since $w_\mu(x_0, \xi_i)$ is a bounded sequence, there is a subsequence, still denoted by ξ_i, such that the limit

$$c_\infty := \lim_{i\to\infty} w_\mu(x_0, \xi_i) \tag{11.3}$$

exists and is not equal to $w_\mu(x_0, \xi)$. Choose $x \in \overline{\mathrm{T}^c(x_0)}$ such that

$$w_\mu(x_0, \xi) = \sup_{g\in \mathrm{T}^c} \langle \mu(gx_0), \xi\rangle = \langle \mu(x), \xi\rangle.$$

Then, for every $i \in \mathbb{N}$, we have

$$\begin{aligned} w_\mu(x_0, \xi) + \langle \mu(x), \xi_i - \xi\rangle &= \langle \mu(x), \xi_i\rangle \\ &\le \sup_{g\in \mathrm{T}^c} \langle \mu(gx_0), \xi_i\rangle \\ &= w_\mu(x_0, \xi_i). \end{aligned}$$

Take the limit $i \to \infty$ to obtain

$$w_\mu(x_0, \xi) \le \lim_{i\to\infty} w_\mu(x_0, \xi_i) = c_\infty$$

and so

$$w_\mu(x_0, \xi) < c_\infty. \tag{11.4}$$

Choose a sequence $x_i \in \overline{\mathrm{T}^c(x_0)}$ such that

$$w_\mu(x_0, \xi_i) = \langle \mu(x_i), \xi_i\rangle \qquad \text{for all } i \in \mathbb{N}. \tag{11.5}$$

Passing to a subsequence we may assume that the limit

$$x_\infty := \lim_{i\to\infty} x_i$$

exists. Then $x_\infty \in \overline{\mathrm{T}^c(x_0)}$ and, by (11.3), (11.4), and (11.5), we have

$$\begin{aligned} w_\mu(x_0, \xi) &< c_\infty \\ &= \lim_{i\to\infty} w_\mu(x_0, \xi_i) \\ &= \lim_{i\to\infty} \langle \mu(x_i), \xi_i \rangle \\ &= \langle \mu(x_\infty), \xi \rangle. \end{aligned}$$

This contradicts equation (11.2) and thus completes the proof of Theorem 11.1. □

The next lemma makes use of the Generalized Kempf Existence Theorem 10.4. Denote the unit sphere in $\mathfrak{t}$ by $S(\mathfrak{t}) := \{\xi \in \mathfrak{t} \mid |\xi| = 1\}$.

Lemma 11.2 *Let $x_0 \in X$ such that $\inf_{g\in\mathrm{T}^c} |\mu(gx_0)| > 0$, and let $\eta \in \overline{\mu(\mathrm{T}^c(x_0))}$, Then the following are equivalent:*

$$|\eta| = \inf_{g\in\mathrm{T}^c} |\mu(gx_0)|, \tag{11.6}$$

$$|\eta|^2 = \inf_{g\in\mathrm{T}^c} \langle \eta, \mu(gx_0) \rangle. \tag{11.7}$$

Moreover, there exists a unique element $\eta \in \overline{\mu(\mathrm{T}^c(x_0))}$ satisfying these conditions, the function $S(\mathfrak{t}) \to \mathbb{R} : \xi \mapsto w_\mu(x_0, \xi)$ takes on its minimum at $\xi_0 := -|\eta|^{-1}\eta$ and only at that point, and

$$\inf_{\xi\in S(\mathfrak{t})} w_\mu(x_0, \xi) = -|\eta| = -\inf_{\mathrm{T}^c(x_0)} |\mu|.$$

Proof Existence is obvious for Eq. (11.6) and uniqueness is obvious for Eq. (11.7). We prove that there exists an element $\eta \in \overline{\mu(\mathrm{T}^c(x_0))}$ that satisfies (11.7). Let $x : \mathbb{R} \to X$ be the solution of (3.2) and define

$$x_\infty := \lim_{t\to\infty} x(t), \qquad \eta := \mu(x_\infty).$$

Then $\eta \in \overline{\mu(\mathrm{T}^c(x_0))}$ and, by Theorem 6.4, we have

$$|\eta| = |\mu(x_\infty)| = \inf_{g\in\mathrm{T}^c} |\mu(gx_0)|.$$

Since T^c is abelian, it follows from (10.8) and (10.9) in Theorem 10.4 that

$$w_\mu(x_0, -\eta) = -|\eta|^2.$$

Hence, by part (i) of Theorem 5.3, we have

$$w_\mu(gx_0, -\eta) = w_\mu(x_0, -\eta) = -|\eta|^2 \qquad \text{for all } g \in \mathrm{T}^c.$$

Since the function $t \mapsto \langle \mu(\exp(-\mathbf{i}t\eta)gx_0), -\eta\rangle$ is nondecreasing by (2.7), and converges to $w_\mu(gx_0, -\eta) = -|\eta|^2$ as t tends to infinity, it follows (by evaluating at $t = 0$) that $\langle \mu(gx_0), -\eta\rangle \leq -|\eta|^2$ and hence

$$|\eta|^2 \leq \langle \mu(gx_0), \eta\rangle$$

for all $g \in \mathrm{T}^c$. This proves the existence of an element $\eta \in \overline{\mu(\mathrm{T}^c(x_0))}$ that satisfies (11.7).

If $\eta \in \overline{\mu(\mathrm{T}^c(x_0))}$ satisfies (11.7) then

$$|\eta|^2 \leq |\eta||\mu(gx_0)|$$

for all $g \in \mathrm{T}^c$ by the Cauchy–Schwarz inequality. Since η is nonzero, this implies $|\eta| \leq |\mu(gx_0)|$ for all $g \in \mathrm{T}^c$. Hence η satisfies (11.6).

Conversely, assume that $\eta \in \overline{\mu(\mathrm{T}^c(x_0))}$ satisfies (11.6). By what we have proved above there is an $\eta_0 \in \overline{\mu(\mathrm{T}^c(x_0))}$ that satisfies (11.7) and hence also (11.6). This implies $|\eta|^2 = |\eta_0|^2 \leq \langle \eta_0, \eta\rangle$ and hence $\eta = \eta_0$. Thus η satisfies (11.7) and is uniquely determined by either condition.

Now define

$$m := \inf_{g \in \mathrm{T}^c} |\mu(gx_0)| > 0$$

and observe that $\langle \mu(gx_0), \xi\rangle \geq -|\mu(gx_0)|$ for all $g \in \mathrm{T}^c$ and all $\xi \in S(\mathfrak{t})$. Take the supremum over all $g \in \mathrm{T}^c$ to obtain

$$w_\mu(x_0, \xi) = \sup_{g \in \mathrm{T}^c} \langle \mu(gx_0), \xi\rangle \geq -\inf_{g \in \mathrm{T}^c} |\mu(gx_0)| = -m$$

for all $\xi \in S(\mathfrak{t})$ by (11.2). Now take the infimum over all $\xi \in S(\mathfrak{t})$ to obtain

$$\inf_{\xi \in S(\mathfrak{t})} w_\mu(x_0, \xi) \geq -m. \tag{11.8}$$

To prove equality, recall from (11.6) and (11.7) that

$$\inf_{g \in \mathrm{T}^c} \langle \mu(gx_0), \eta\rangle = |\eta|^2 = m^2. \tag{11.9}$$

Thus $\eta \neq 0$ and, by Eq. (11.2), we have

$$w_\mu\left(x_0, -|\eta|^{-1}\eta\right) = -\inf_{g \in \mathrm{T}^c} \langle \mu(gx_0), |\eta|^{-1}\eta\rangle = -|\eta| = -m.$$

By (11.8) this shows that the function $S(\mathfrak{t}) \to \mathbb{R} : \xi \mapsto w_\mu(x_0, \xi)$ attains its minimum at the point $\xi_0 := -|\eta|^{-1}\eta \in S(\mathfrak{t})$, i.e.

$$\inf_{\xi \in S(\mathfrak{t})} w_\mu(x_0, \xi) = w_\mu\left(x_0, -|\eta|^{-1}\eta\right) = -m.$$

Now let $\xi \in S(\mathfrak{t})$ such that $\xi \neq -|\eta|^{-1}\eta$. Then $\langle |\eta|^{-1}\eta, \xi\rangle > -1$ and hence

$$\begin{aligned} w_\mu(x_0, \xi) &= \sup_{g \in \mathrm{T}^c} \langle \mu(gx_0), \xi\rangle \\ &\geq \langle \eta, \xi\rangle \\ &> -|\eta| \\ &= -m. \end{aligned}$$

Here the first step follows from (11.2), the second step follows from the fact that $\eta \in \overline{\mu(\mathrm{T}^c(x_0))}$, the third step uses the inequality $\langle |\eta|^{-1}\eta, \xi\rangle > -1$, and the last step follows from (11.9). This proves Lemma 11.2. □

The following theorem asserts that the image of every complexified group orbit under the moment map has a convex closure. This is a variant of Atiyah–Guillemin–Sternberg convexity (see [1] and [66, §5 & Appendix]).

Theorem 11.3 (Convexity) *For every $x \in X$ the set $\overline{\mu(\mathrm{T}^c(x))}$ is convex.*

Proof For every $\tau \in \mathfrak{t}$ and every $r > 0$ denote the closed ball of radius r about τ by $B_r(\tau) \subset \mathfrak{t}$. Fix an element $x \in X$ and define

$$\Delta := \overline{\mu(\mathrm{T}^c(x))}.$$

For $\tau \in \mathfrak{t} \setminus \Delta$ define

$$d(\tau, \Delta) := \inf_{\xi \in \Delta} |\tau - \xi|.$$

Then Lemma 11.2, with μ replaced by $\mu - \tau$, asserts that

$$\left.\begin{array}{l} \tau \in \mathfrak{t} \setminus \Delta \text{ and} \\ \eta \in B_{d(\tau,\Delta)}(\tau) \cap \Delta \end{array}\right\} \quad \Longrightarrow \quad d(\tau, \Delta)^2 = \inf_{\eta' \in \Delta} \langle \eta - \tau, \eta' - \tau\rangle. \tag{11.10}$$

This implies that Δ is convex. To see this, suppose by contradiction that there exist elements $\tau_0, \tau_1 \in \Delta$ and a constant $0 < \lambda < 1$ such that $\tau := (1 - \lambda)\tau_0 + \lambda\tau_1 \notin \Delta$. Then $r := d(\tau, \Delta) > 0$. Fix an element $\eta \in B_r(\tau) \cap \Delta$. Then by (11.10) we have

$$\langle \eta - \tau, \tau_i - \tau\rangle \geq r^2 \qquad \text{for } i = 0, 1.$$

However, since $\tau_0 - \tau = \lambda(\tau_0 - \tau_1)$ and $\tau_1 - \tau = (1 - \lambda)(\tau_1 - \tau_0)$, the inner products $\langle \eta - \tau, \tau_0 - \tau \rangle$ and $\langle \eta - \tau, \tau_1 - \tau \rangle$ have opposite signs. This is a contradiction and proves Theorem 11.3. □

Chapter 12
The Hilbert–Mumford Criterion

In this chapter we return to the general case where G is any compact Lie group acting on a closed Kähler manifold X by Kähler isometries, and the action is generated by an equivariant moment map $\mu : X \to \mathfrak{g} = \mathrm{Lie}(\mathrm{G})$.

Theorem 12.1 (The Mumford Numerical Function)

(i) *For every $x \in X$ we have*

$$\begin{aligned} m_\mu(x) &:= \inf_{0 \neq \xi \in \mathfrak{g}} \frac{w_\mu(x,\xi)}{|\xi|} \\ &= \inf_{\zeta \in \mathscr{T}^c} \frac{w_\mu(x,\zeta)}{\sqrt{|\mathrm{Re}(\zeta)|^2 - |\mathrm{Im}(\zeta)|^2}} \\ &= \inf_{\xi \in \Lambda} \frac{w_\mu(x,\xi)}{|\xi|} \\ &= \inf_{\zeta \in \Lambda^c} \frac{w_\mu(x,\zeta)}{\sqrt{|\mathrm{Re}(\zeta)|^2 - |\mathrm{Im}(\zeta)|^2}}. \end{aligned} \tag{12.1}$$

The function $m_\mu : X \to \mathbb{R}$ defined by Eq. (12.1) *for $x \in X$ is called the* **Mumford numerical function**.

(ii) *The Mumford numerical function $m_\mu : X \to \mathbb{R}$ is* G^c*-invariant.*

(iii) *Every $x \in X$ satisfies $m_\mu(x) + \inf_{\mathrm{G}^c(x)} |\mu| \geq 0$.*

(iv) *If $x \in X$ is μ-unstable then* $0 > m_\mu(x) = -\inf_{\mathrm{G}^c(x)} |\mu|$.

(v) *If $x \in X$ is μ-stable then* $0 < m_\mu(x) \leq \inf_{\overline{\mathrm{G}^c(x)} \setminus \mathrm{G}^c(x)} |\mu|$.

(vi) *For each $x \in X$ there is a $\xi \in \mathfrak{g}$ such that* $|\xi| = 1$ *and* $w_\mu(x,\xi) = m_\mu(x)$.

Proof By Theorem D.4 every $\zeta \in \mathscr{T}^c$ is equivalent to an element $\xi \in \mathfrak{g}$, and we have $w_\mu(x,\zeta) = w_\mu(x,\xi)$ by Theorem 5.3 (ii) and $|\mathrm{Re}(\zeta)|^2 - |\mathrm{Im}(\zeta)|^2 = |\xi|^2$ by Lemma 5.7. This proves the second and last equalities in (12.1).

© The Author(s), under exclusive license to Springer Nature Switzerland AG 2021
V. Georgoulas et al., *The Moment-Weight Inequality and the Hilbert–Mumford Criterion*, Lecture Notes in Mathematics 2297,
https://doi.org/10.1007/978-3-030-89300-2_12

It remains to prove that the infimum over $\mathfrak{g}\setminus\{0\}$ in (12.1) agrees with the infimum over Λ. Let $x_0 \in X$ and $\xi_0 \in \mathfrak{g}$ such that

$$|\xi_0| = 1, \qquad w_\mu(x_0, \xi_0) = m_\mu(x_0)$$

(Theorem 10.3). Consider the torus

$$\mathrm{T} := \overline{\{\exp(t\xi_0) \mid t \in \mathbb{R}\}} \subset \mathrm{G}, \qquad \mathfrak{t} := \mathrm{Lie}(\mathrm{T}).$$

Denote the unit sphere in $\mathfrak{t}$ by

$$S(\mathfrak{t}) := \{\xi \in \mathfrak{t} \mid |\xi| = 1\}.$$

Then the function $S(\mathfrak{t}) \to \mathbb{R} : \xi \mapsto w_\mu(x_0, \xi)$ is continuous by Theorem 11.1, the set $\{|\eta|^{-1}\eta \mid \eta \in \mathfrak{t} \cap \Lambda\}$ is dense in $S(\mathfrak{t})$, and

$$\inf_{\xi \in S(\mathfrak{t})} w_\mu(x_0, \xi) = m_\mu(x_0).$$

Hence, for each $\varepsilon > 0$, there exists an element $\eta \in \mathfrak{t} \cap \Lambda$ such that

$$|\eta|^{-1} w_\mu(x_0, \eta) = w_\mu(x_0, |\eta|^{-1}\eta) < m_\mu(x_0) + \varepsilon.$$

This implies

$$\inf_{\eta \in \Lambda} (|\eta|^{-1} w_\mu(x_0, \eta)) \leq m_\mu(x_0) = \inf_{0 \neq \xi \in \mathfrak{g}} (|\xi|^{-1} w_\mu(x_0, \xi)).$$

The converse inequality is obvious and this proves (i). Part (ii) follows from (i), Theorem 5.3, and Lemma 5.7, part (iii) is equivalent to the moment-weight inequality in Theorem 6.7, part (iv) follows from Corollary 10.6, part (v) follows from Theorem 10.3, and part (vi) from Theorem 10.1. This proves Theorem 12.1. □

The main results of the present chapter are the Hilbert–Mumford numerical criteria for μ-semistability, μ-polystability, and μ-stability. We begin with the μ-semistable case, where the Hilbert–Mumford criterion is a direct consequence of Theorems 7.4 (necessity) and 12.1 (sufficiency).

Theorem 12.2 (Hilbert–Mumford Criterion: Semistable Case) *For every $x_0 \in X$ the following are equivalent.*

(i) *x_0 is μ-semistable.*
(ii) *Every $\xi \in \Lambda$ satisfies $w_\mu(x_0, \xi) \geq 0$.*
(iii) *Every $\xi \in \mathfrak{g} \setminus \{0\}$ satisfies $w_\mu(x_0, \xi) \geq 0$.*
(iv) *Every $\zeta \in \Lambda^c$ satisfies $w_\mu(x_0, \zeta) \geq 0$.*
(v) *Every $\zeta \in \mathscr{T}^c$ satisfies $w_\mu(x_0, \zeta) \geq 0$.*

Proof The equivalence of the assertions (ii), (iii), (iv) and (v) follows from Eq. (12.1) in Theorem 12.1, that each of these conditions implies (i) follows from part (iv) of Theorem 12.1, and that (i) implies (v) was proved in part (i) of Theorem 7.4.

More precisely, Mumford's Theorems D.4 and 5.3 show that (ii) $\iff$ (iv) and (iii) $\iff$ (v). The equivalence (ii) $\iff$ (iii) follows from the continuity of the weights for torus actions in Theorem 11.1, with the argument spelled out in the proof of Theorem 12.1. That (i) $\implies$ (iii) follows from the moment-weight inequality in Theorem 6.7, which shows that the existence of a negative weight implies that x_0 is μ-unstable. That (iii) $\implies$ (i) follows from the Kempf Existence Theorem 10.4, which produces an element $\xi \in \mathfrak{g} \setminus \{0\}$ with $w_\mu(x_0, \xi) < 0$ whenever x_0 is μ-unstable. This proves Theorem 12.2. □

Theorem 12.3 (Hilbert–Mumford Criterion: Unstable Case) *For every $x_0 \in X$ the following are equivalent.*

(i) *x_0 is μ-unstable.*
(ii) *There exists a $\xi \in \Lambda$ such that $w_\mu(x_0, \xi) < 0$.*
(iii) *There exists a $\xi \in \mathfrak{g} \setminus \{0\}$ such that $w_\mu(x_0, \xi) < 0$.*
(iv) *There exists a $\zeta \in \Lambda^c$ such that $w_\mu(x_0, \zeta) < 0$.*
(v) *There exists a $\zeta \in \mathscr{T}^c$ such that $w_\mu(x_0, \zeta) < 0$.*

Proof This follows directly from the definitions and Theorem 12.2. □

The next theorem is the **Hilbert–Mumford numerical criterion** in its classical form. We derive it as a corollary of Theorem 12.3.

Theorem 12.4 (Hilbert–Mumford Criterion: Classical Case) *Let $\mathrm{G} \subset \mathrm{U}(n)$ be a compact Lie group, let $\mathrm{G}^c \subset \mathrm{GL}(n, \mathbb{C})$ be its complexification, and let V be a finite-dimensional complex vector space equipped with a holomorphic representation of G^c. If $v \in V$ is a nonzero vector such that $0 \in \overline{\mathrm{G}^c(v)}$, then there exists an element $\xi \in \Lambda$ such that $\lim_{t\to\infty} \exp(\mathbf{i}t\xi)v = 0$.*

Proof By assumption G induces a Hamiltonian group action on the projective space

$$X = \mathbb{P}(V)$$

with the moment map of Lemma 8.2. Let $v \in V$ be a nonzero vector such that

$$0 \in \overline{\mathrm{G}^c(v)}.$$

Then v is unstable and thus the point $x := [v] \in \mathbb{P}(V)$ is μ-unstable by Theorem 8.5. Hence Theorem 12.3 asserts that there exists a $\xi \in \Lambda$ such that $w_\mu(x, \xi) < 0$. By Lemma 8.4 this means that v is contained in the direct sum of the negative eigenspaces of the Hermitian operator on V determined by $\mathbf{i}\xi$. Hence

$$\lim_{t\to\infty} \exp(\mathbf{i}t\xi)v = 0$$

and this proves Theorem 12.4. □

Theorem 12.5 (Hilbert–Mumford Criterion: Polystable Case) *For every $x_0 \in X$ the following are equivalent.*

(i) *x_0 is μ-polystable.*
(ii) *Every $\xi \in \Lambda$ satisfies $w_\mu(x_0, \xi) \geq 0$ and*

$$w_\mu(x_0, \xi) = 0 \quad \Longrightarrow \quad \lim_{t \to \infty} \exp(\mathbf{i}t\xi)x_0 \in \mathrm{G}^c(x_0).$$

(iii) *Every $\xi \in \mathfrak{g} \setminus \{0\}$ satisfies $w_\mu(x_0, \xi) \geq 0$ and*

$$w_\mu(x_0, \xi) = 0 \quad \Longrightarrow \quad \lim_{t \to \infty} \exp(\mathbf{i}t\xi)x_0 \in \mathrm{G}^c(x_0).$$

(iv) *Every $\zeta \in \Lambda^c$ satisfies $w_\mu(x_0, \zeta) \geq 0$ and*

$$w_\mu(x_0, \zeta) = 0 \quad \Longrightarrow \quad \lim_{t \to \infty} \exp(\mathbf{i}t\zeta)x_0 \in \mathrm{G}^c(x_0).$$

(v) *Every $\zeta \in \mathscr{T}^c$ satisfies $w_\mu(x_0, \zeta) \geq 0$ and*

$$w_\mu(x_0, \zeta) = 0 \quad \Longrightarrow \quad \lim_{t \to \infty} \exp(\mathbf{i}t\zeta)x_0 \in \mathrm{G}^c(x_0).$$

Proof That (i) $\Longrightarrow$ (v) was proved in part (ii) of Theorem 7.4, and the implications (v) $\Longrightarrow$ (iv) $\Longrightarrow$ (ii) and (v) $\Longrightarrow$ (iii) $\Longrightarrow$ (ii) follow directly from the definitions. Thus it remains to prove that (ii) implies (i).

The proof of sufficiency of the μ-weight condition for μ-polystability is due to Chen–Sun [23, Theorem 4.7]. Here is their argument. Fix an element $x_0 \in X$ that is μ-semistable but not μ-polystable, let $x : \mathbb{R} \to X$ be the solution of the differential equation

$$\dot{x} = -JL_x\mu(x), \qquad x(0) = x_0,$$

and define

$$x_\infty := \lim_{t \to \infty} x(t).$$

Then, by Theorem 7.2,

$$x_\infty := \lim_{t \to \infty} x(t) \notin \mathrm{G}^c(x_0), \qquad \mu(x_\infty) = 0, \qquad \ker L_{x_\infty} \neq 0. \tag{12.2}$$

We prove in seven steps that there exists an element $\xi \in \Lambda$ such that

$$w_\mu(x_0, \xi) = 0, \qquad \lim_{t \to \infty} \exp(\mathbf{i}t\xi)x_0 \notin \mathrm{G}^c(x_0). \tag{12.3}$$

Step 1 *The complex isotropy subgroup* $\mathrm{G}^c_{x_\infty}$ *is the complexification of* G_{x_∞}. *Moreover, there exists a* G_{x_∞}*-equivariant local holomorphic coordinate chart* $\psi : U_\infty \to X$ *on a* G_{x_∞}*-invariant open neighborhood* $U_\infty \subset T_{x_\infty}X$ *of the origin such that*

$$\psi(0) = x_\infty, \qquad d\psi(0) = \mathrm{id}.$$

Since $\mu(x_\infty) = 0$, it follows from Lemma 2.3 that $\mathrm{G}^c_{x_\infty}$ is the complexification of G_{x_∞}. Now let $\phi : (T_{x_\infty}X, 0) \to (X, x_\infty)$ be any holomorphic coordinate chart, defined in a neighborhood of the origin in $T_{x_\infty}X$, such that $\phi(0) = x_\infty$ and $d\phi(0) = \mathrm{id}$. Let dvol_∞ denote the Haar measure on G_{x_∞} and define a map χ from an open neighborhood of x_∞ in X to an open neighborhood of the origin in $T_{x_\infty}X$ by

$$\chi(x) := \frac{1}{\mathrm{Vol}(\mathrm{G}_{x_\infty})} \int_{\mathrm{G}_{x_\infty}} u^{-1}\phi^{-1}(ux)\mathrm{dvol}_\infty(u)$$

for $x \in X$ sufficiently close to x_∞. Then $\chi(x_\infty) = 0$, $d\chi(x_\infty) = \mathrm{id}$, and χ is holomorphic and G_{x_∞}-equivariant. Hence, by the inverse function theorem, it restricts to a G_{x_∞}-equivariant holomorphic diffeomorphism from a G_{x_∞}-invariant open neighborhood of x_∞ in X to a G_{x_∞}-invariant open neighborhood $U_\infty \subset T_{x_\infty}X$ of the origin. The inverse $\psi := \chi^{-1}$ of this restriction satisfies the requirements of Step 1.

Step 2 *Recall the notation* $v_\xi(x) = L_x\xi$ *for the infinitesimal action of* $\xi \in \mathfrak{g}$ *on* X, *and let* $\psi : U_\infty \to X$ *be the as in Step 1. Then there exists a constant* $\delta > 0$ *such that*

$$B_\delta(x_\infty) := \left\{\widehat{x} \in T_{x_\infty}X \,\middle|\, |\widehat{x}| < \delta\right\} \subset U_\infty$$

and, for all $\widehat{x} \in B_\delta(x_\infty) \cap \mathrm{im}\,(L^c_{x_\infty})^\perp$ *and all* $\widehat{y} \in T_{x_\infty}X$, $\zeta = \xi + \mathbf{i}\eta \in \mathfrak{g}^c$,

$$\begin{array}{l} d\psi(\widehat{x})\widehat{y} = L^c_{\psi(\widehat{x})}\zeta \\ \text{and } \widehat{y} \perp \mathrm{im}\,(L^c_{x_\infty}) \end{array} \iff \begin{array}{l} \widehat{y} = \nabla_{\widehat{x}} v_\xi(x_\infty) + J\nabla_{\widehat{x}} v_\eta(x_\infty) \\ \text{and } L^c_{x_\infty}\zeta = 0. \end{array} \tag{12.4}$$

The action of G_{x_∞} on X by Kähler isometries gives rise to a unitary action of G_{x_∞} on the tangent space $T_{x_\infty}X$ at the fixed point x_∞. The infinitesimal action takes the form of a Lie algebra homomorphism

$$\mathfrak{g}_{x_\infty} := \mathrm{Lie}(\mathrm{G}_{x_\infty}) = \ker(L_{x_\infty}) \to \mathfrak{u}(T_{x_\infty}X) : \xi \mapsto A_\xi = \nabla v_\xi(x_\infty).$$

Thus

$$A_\xi\widehat{x} := \tfrac{d}{dt}\Big|_{t=0} \exp(t\xi)\widehat{x} = \nabla_{\widehat{x}} v_\xi(x_\infty)$$

for $\widehat{x} \in T_{x_\infty} X$ and $\xi \in \ker(L_{x_\infty})$. Differentiate the identity $\psi(u\widehat{x}) = u\psi(\widehat{x})$ for $\widehat{x} \in U_\infty$ and $u \in \mathrm{G}_{x_\infty}$, to obtain for all $\widehat{x} \in U_\infty$ and all $\xi \in \mathfrak{g}$,

$$L_{x_\infty}\xi = 0 \quad \Longrightarrow \quad d\psi(\widehat{x})A_\xi\widehat{x} = L_{\psi(\widehat{x})}\xi. \tag{12.5}$$

Since $uL_{x_\infty}\eta = L_{x_\infty}(u\eta u^{-1})$ for $u \in \mathrm{G}_{x_\infty}$ and $\eta \in \mathfrak{g}$, we also have for $\xi, \eta \in \mathfrak{g}$,

$$L_{x_\infty}\xi = 0 \quad \Longrightarrow \quad A_\xi L_{x_\infty}\eta = L_{x_\infty}[\xi, \eta]. \tag{12.6}$$

By (12.6) the subspaces $\mathrm{im}\,(L^c_{x_\infty})$ and $\mathrm{im}\,(L^c_{x_\infty})^\perp$ are invariant under A_ξ for every $\xi \in \ker(L_{x_\infty})$. Hence the implication "$\Longleftarrow$" in (12.4) for all $\widehat{x} \in U_\infty \cap \mathrm{im}\,(L^c_{x_\infty})^\perp$ follows from (12.5), because the derivative $d\psi(\widehat{x})$ is complex linear and $L^c_{x_\infty}\zeta = 0$ if and only if $L_{x_\infty}\xi = L_{x_\infty}\eta = 0$.

For $\widehat{x} \in U_\infty$ define the linear operator $\mathscr{L}_{\widehat{x}} : \mathfrak{g}^c \times (\mathrm{im}\,L^c_{x_\infty})^\perp \to T_{\psi(\widehat{x})}X$ by

$$\mathscr{L}_{\widehat{x}}(\zeta, \widehat{y}) := L^c_{\psi(\widehat{x})}\zeta - d\psi(\widehat{x})\widehat{y}.$$

for $\zeta \in \mathfrak{g}^c$ and $\widehat{y} \in \mathrm{im}\,(L^c_{x_\infty})^\perp$. The index of this operator (the dimension of the source minus the dimension of the target) is $2k$, where $k := \dim(\mathrm{G}_{x_\infty})$. Moreover, if $\widehat{x} \in U_\infty \cap \mathrm{im}\,(L^c_{x_\infty})^\perp$, then it follows from the implication "$\Longleftarrow$" in (12.4) (already proved) that the $2k$-dimensional subspace

$$\mathscr{Z}_{\widehat{x}} := \left\{(\xi + \mathbf{i}\eta,\, A_\xi\widehat{x} + JA_\eta\widehat{x}) \,\middle|\, \xi, \eta \in \ker(L_{x_\infty})\right\} \subset \mathfrak{g}^c \times \mathrm{im}\,(L^c_{x_\infty})^\perp$$

is contained in the kernel of the operator $\mathscr{L}_{\widehat{x}}$. Since $\mathscr{L}_{\widehat{x}}$ is surjective for $\widehat{x} = 0$, there exists a constant $\delta > 0$ such that $B_\delta(x_\infty) \subset U_\infty$ and the operator $\mathscr{L}_{\widehat{x}}$ is surjective for every $\widehat{x} \in B_\delta(x_\infty)$. Hence $\dim(\ker(\mathscr{L}_{\widehat{x}})) = 2k$ and so $\ker(\mathscr{L}_{\widehat{x}}) = \mathscr{Z}_{\widehat{x}}$ for every $\widehat{x} \in B_\delta(x_\infty) \cap \mathrm{im}\,(L^c_{x_\infty})^\perp$. This proves Step 2.

Step 3 *Let $\delta > 0$ be the constant in Step 2. Then there exists a $t_0 > 0$ and smooth curves $\xi, \eta : [t_0, \infty) \to (\ker L_{x_\infty})^\perp$ and $\widehat{x} : [t_0, \infty) \to (\mathrm{im}\,L^c_{x_\infty})^\perp$ such that*

$$x(t) = \exp(\mathbf{i}\eta(t))\exp(\xi(t))\psi(\widehat{x}(t)), \qquad |\widehat{x}(t)| < \delta,$$

for every $t \geq t_0$.

Define the map $f : \ker(L_{x_\infty})^\perp \times \ker(L_{x_\infty})^\perp \times (B_\delta(x_\infty) \cap \mathrm{im}\,(L^c_{x_\infty})^\perp) \to X$ by

$$f(\xi, \eta, \widehat{x}) := \exp(\mathbf{i}\eta)\exp(\xi)\psi(\widehat{x}).$$

Its derivative at the origin is bijective. Hence f restricts to a diffeomorphism from an open neighborhood of the origin in $\ker(L_{x_\infty})^\perp \times \ker(L_{x_\infty})^\perp \times \mathrm{im}\,(L^c_{x_\infty})^\perp$ onto an open neighborhood of x_∞ in X. This proves Step 3.

Step 4 *Let $t_0, \xi, \eta, \widehat{x}$ be as in Step 3 and let $g : \mathbb{R} \to G^c$ be the unique solution of the initial value problem $g^{-1}\dot{g} = \mathbf{i}\mu(x)$, $g(0) = \mathbb{1}$. For $t \geq t_0$ define*

$$h(t) := \exp(\mathbf{i}\eta(t))\exp(\xi(t)), \qquad g_\infty(t) := h(t_0)^{-1}g(t_0)^{-1}g(t)h(t).$$

Then, for all $t \geq t_0$,

$$g_\infty(t_0) = \mathbb{1}, \quad g_\infty(t) \in G^c_{x_\infty}, \quad \widehat{x}(t) = g_\infty(t)^{-1}\widehat{x}(t_0), \quad x(t) = h(t)\psi(\widehat{x}(t)).$$

By Lemma 3.2 and Step 3, we have

$$g(t)^{-1}x_0 = x(t) = h(t)\psi(\widehat{x}(t))$$

for $t \geq t_0$. Hence, for all $t \geq t_0$,

$$\begin{aligned}
\psi(\widehat{x}(t)) &= h(t)^{-1}g(t)^{-1}x_0 \\
&= h(t)^{-1}g(t)^{-1}g(t_0)h(t_0)\psi(\widehat{x}(t_0)) \\
&= g_\infty(t)^{-1}\psi(\widehat{x}(t_0)).
\end{aligned}$$

Differentiate this equation to obtain

$$d\psi(\widehat{x})\partial_t\widehat{x} = L^c_{\psi(\widehat{x})}\zeta_\infty, \qquad \zeta_\infty := \xi_\infty + \mathbf{i}\eta_\infty := -g_\infty^{-1}\dot{g}_\infty.$$

Since $|\widehat{x}(t)| < \delta$ and $\partial_t\widehat{x}(t) \perp \operatorname{im}(L^c_{x_\infty})$, it follows from Step 2 that

$$\zeta_\infty(t) \in \ker(L^c_{x_\infty}), \qquad \partial_t\widehat{x}(t) = \nabla_{\widehat{x}(t)}v_{\xi_\infty(t)}(x_\infty) + J\nabla_{\widehat{x}(t)}v_{\eta_\infty(t)}(x_\infty)$$

for $t \geq t_0$. Since $g_\infty(t_0) = \mathbb{1}$, we obtain

$$g_\infty(t) \in G^c_{x_\infty}, \qquad \widehat{x}(t) = g_\infty(t)^{-1}\widehat{x}(t_0)$$

for $t \geq t_0$. This proves Step 4.

Step 5 *There exists an element $\xi \in \Lambda$ such that $L_{x_\infty}\xi = 0$ and*

$$\lim_{t\to\infty} \exp(\mathbf{i}t\xi)h(t_0)^{-1}x(t_0) = x_\infty, \qquad w_\mu(h(t_0)^{-1}x(t_0), \xi) = 0.$$

By Step 4, we have

$$\widehat{x}(t) \in G^c_{x_\infty}(\widehat{x}(t_0))$$

for every $t \geq t_0$ and

$$\lim_{t\to\infty} \widehat{x}(t) = 0.$$

Moreover, the compact Lie group G_{x_∞} acts on $\mathrm{im}\,(L^c_{x_\infty})^\perp$ by unitary automorphisms. Hence, by Theorem 12.4, there exists an element $\xi \in \Lambda$ such that

$$\lim_{t\to\infty} \exp(\mathbf{i}t\xi)\widehat{x}(t_0) = 0, \qquad L_{x_\infty}\xi = 0.$$

Since $\mathbf{i}\xi$ acts on the tangent space $T_{x_\infty}X$ by a Hermitian endomorphism, it follows that the function

$$t \mapsto |\exp(\mathbf{i}t\xi)\widehat{x}(t_0)|$$

is decreasing. Hence the vector

$$\exp(\mathbf{i}t\xi)\widehat{x}(t_0) \in \mathrm{im}\,(L^c_{x_\infty})^\perp$$

is contained in the domain U_∞ of the holomorphic coordinate chart ψ for all $t \geq 0$. Hence

$$\psi(\exp(\mathbf{i}t\xi)\widehat{x}(t_0)) = \exp(\mathbf{i}t\xi)\psi(\widehat{x}(t_0)) = \exp(\mathbf{i}t\xi)h(t_0)^{-1}x(t_0)$$

for $t \geq 0$, hence

$$\lim_{t\to\infty} \exp(\mathbf{i}t\xi)h(t_0)^{-1}x(t_0) = \psi(0) = x_\infty$$

and hence

$$w_\mu(h(t_0)^{-1}x(t_0), \xi) = \langle \mu(x_\infty), \xi \rangle = 0.$$

This proves Step 5.

Step 6 *There exists an element $\zeta \in \Lambda^c$ such that*

$$w_\mu(x_0, \zeta) = 0, \qquad \lim_{t\to\infty} \exp(\mathbf{i}t\zeta)x_0 \notin \mathrm{G}^c(x_0).$$

Let g, h, ξ be as in Steps 4 and 5 and define

$$\zeta := g(t_0)h(t_0)\xi h(t_0)^{-1}g(t_0)^{-1}.$$

Then, by part (i) of Theorem 5.3,

$$\begin{aligned} w_\mu(x_0, \zeta) &= w_\mu(h(t_0)^{-1}g(t_0)^{-1}x_0, h(t_0)^{-1}g(t_0)^{-1}\zeta g(t_0)h(t_0)) \\ &= w_\mu(h(t_0)^{-1}x(t_0), \xi) \\ &= 0 \end{aligned}$$

and

$$\begin{aligned} \lim_{t\to\infty} \exp(\mathbf{i}t\zeta)x_0 &= \lim_{t\to\infty} \exp\left(\mathbf{i}t g(t_0)h(t_0)\xi h(t_0)^{-1}g(t_0)^{-1}\right)x_0 \\ &= g(t_0)h(t_0) \lim_{t\to\infty} \exp(\mathbf{i}t\xi)h(t_0)^{-1}x(t_0) \\ &= g(t_0)h(t_0)x_\infty. \end{aligned}$$

Hence $\lim_{t\to\infty} \exp(\mathbf{i}t\zeta)x_0 \notin \mathrm{G}^c(x_0)$ by (12.2) and this proves Step 6.

Step 7 *There exists an element $\xi \in \Lambda$ that satisfies* (12.3)*, i.e.*

$$w_\mu(x_0, \xi) = 0, \qquad \lim_{t\to\infty} \exp(\mathbf{i}t\xi)x_0 \notin \mathrm{G}^c(x_0).$$

Let $\zeta \in \Lambda^c$ be as in Step 6. By Theorem D.4 there exist elements

$$p, p^+ \in \mathrm{P}(\zeta)$$

such that

$$\xi := p^{-1}\zeta p \in \Lambda, \qquad p^+ = \lim_{t\to\infty} \exp(\mathbf{i}t\zeta)p\exp(-\mathbf{i}t\zeta).$$

Hence, by part (ii) of Theorem 5.3,

$$w_\mu(x_0, \xi) = w_\mu(x_0, \zeta) = 0$$

and

$$\begin{aligned} \lim_{t\to\infty} \exp(\mathbf{i}t\xi)x_0 &= \lim_{t\to\infty} p^{-1}\exp(\mathbf{i}t\zeta)p\exp(-\mathbf{i}t\zeta)\exp(\mathbf{i}t\zeta)x_0 \\ &= p^{-1}p^+ \lim_{t\to\infty} \exp(\mathbf{i}t\zeta)x_0 \\ &\notin \mathrm{G}^c(x_0). \end{aligned}$$

The last assertion follows from Step 6. This proves Step 7 and Theorem 12.5.

□

Theorem 12.6 (Hilbert–Mumford Criterion: Stable Case) *For every $x_0 \in X$ the following are equivalent.*

(i) *x_0 is μ-stable.*
(ii) *Every $\xi \in \Lambda$ satisfies $w_\mu(x_0, \xi) > 0$.*
(iii) *Every $\xi \in \mathfrak{g} \setminus \{0\}$ satisfies $w_\mu(x_0, \xi) > 0$.*
(iv) *Every $\zeta \in \Lambda^c$ satisfies $w_\mu(x_0, \zeta) > 0$.*
(v) *Every $\zeta \in \mathscr{T}^c$ satisfies $w_\mu(x_0, \zeta) > 0$.*

Proof That (i) $\Longrightarrow$ (v) was proved in part (iii) of Theorem 7.4, and the implications (v) $\Longrightarrow$ (iv) $\Longrightarrow$ (ii) and (v) $\Longrightarrow$ (iii) $\Longrightarrow$ (ii) follow directly from the definitions. Thus it remains to prove that (ii) implies (i).

Assume $w_\mu(x_0, \xi) > 0$ for all $\xi \in \Lambda$. Then x_0 is μ-polystable by Theorem 12.5. Hence there exists an element $g \in \mathrm{G}^c$ such that $\mu(gx_0) = 0$. Assume, by contradiction, that $\ker L_{gx_0} \neq \{0\}$. Then the isotropy subgroup G_{gx_0} is not discrete and hence $\Lambda \cap \ker L_{gx_0} \neq \emptyset$. Fix any element $\xi_0 \in \Lambda \cap \ker L_{gx_0}$. Then $w_\mu(gx_0, \xi_0) = 0$ by Lemma 7.5, and this implies $w_\mu(x_0, g^{-1}\xi_0 g) = 0$ by part (i) of Theorem 5.3. Since $g^{-1}\xi_0 g \in \Lambda^c$, it follows from Theorem D.4 and part (ii) of Theorem 5.3 that there exists an element $\xi \in \Lambda$ such that $w_\mu(x_0, \xi) = 0$, in contradiction to our assumption. This contradiction shows that $\ker L_{gx_0} = \{0\}$, hence $\ker L^c_{gx_0} = \{0\}$ by Lemma 2.2, and therefore $\ker L^c_{x_0} = \{0\}$. Thus x_0 is μ-stable and this proves Theorem 12.6. □

Corollary 12.7 *For every $x_0 \in X$ the following holds.*

(i) *x_0 is μ-unstable if and only if $m_\mu(x_0) < 0$.*
(ii) *x_0 is μ-semistable if and only if $m_\mu(x_0) \geq 0$.*
(iii) *x_0 is μ-stable if and only if $m_\mu(x_0) > 0$.*
(iv) *x_0 is μ-stable if and only if the moment-weight inequality* (10.1) *is strict, i.e. $m_\mu(x_0) + \inf_{\mathrm{G}^c(x_0)} |\mu| > 0$.*

Proof Part (i) follows directly from Theorem 12.3 and the definition of the Mumford numerical function $m_\mu : X \to \mathbb{R}$ in (12.1). Part (ii) follows from (i). Part (iii) follows from Theorems 12.6 and 10.3. To prove part (iv), observe that

$$m_\mu(x_0) > 0 = \inf_{\mathrm{G}^c(x_0)} |\mu|$$

whenever x_0 is μ-stable (by part (iii)), that

$$-m_\mu(x_0) = \inf_{\mathrm{G}^c(x_0)} |\mu|$$

whenever x_0 is μ-unstable (by Corollary 10.6), and that

$$m_\mu(x_0) = 0 = \inf_{\mathrm{G}^c(x_0)} |\mu|$$

whenever x_0 is μ-semistable, but not μ-stable (by parts (ii) and (iii)). This proves Corollary 12.7. □

Chapter 13
Critical Orbits

The results of this chapter are based on the work of Gábor Székelyhidi [72]. We assume throughout that G is a compact Lie group whose Lie algebra $\mathfrak{g} = \mathrm{Lie}(\mathrm{G})$ is equipped with an invariant inner product, that it acts on a closed Kähler manifold X by Kähler isometries, and that the action is generated by an equivariant moment map $\mu : X \to \mathfrak{g}$. The goal of this section is to examine complexified group orbits that contain critical points of the moment map squared. In the notation of Chap. 2 the problem can be rephrased as that of finding a solution of the equation

$$L_x\mu(x) = 0$$

as x ranges over a G^c-orbit in X. This is the finite-dimensional analogue of finding an *extremal metric* in the cscK setting described in Chap. 1. The main result is the generalized Székelyhidi criterion in Theorem 13.2 for the existence of a critical point in the complexified group orbit, in terms of polystability with respect to the action of a suitable quotient group. As a warmup we begin with the following criterion in terms of the gradient flow of the square of the moment map, which is analogous to Theorem 7.2.

Theorem 13.1 (Critical Orbits) *Let $x_0 \in X$, let $x : \mathbb{R} \to X$ be the unique solution of* (3.2)*, and define $x_\infty := \lim_{t\to\infty} x(t)$. Then the following are equivalent.*

(i) *$\mathrm{G}^c(x_0)$ contains a critical point of the square of the moment map.*
(ii) *$x_\infty \in \mathrm{G}^c(x_0)$.*

Proof We prove that (i) implies (ii). Assume that there exists an $x \in \mathrm{G}^c(x_0)$ such that $L_x\mu(x) = 0$. Then $x, x_\infty \in \overline{\mathrm{G}^c(x_0)}$ and

$$|\mu(x)| = \inf_{g\in\mathrm{G}^c} \mu(gx_0) = |\mu(x_\infty)|$$

© The Author(s), under exclusive license to Springer Nature Switzerland AG 2021
V. Georgoulas et al., *The Moment-Weight Inequality and the Hilbert–Mumford Criterion*, Lecture Notes in Mathematics 2297,
https://doi.org/10.1007/978-3-030-89300-2_13

by Corollary 6.2 and Theorem 6.4. Hence $x_\infty \in \mathrm{G}(x) \subset \mathrm{G}^c(x_0)$ by Theorem 6.5. This shows that (i) implies (ii). The converse implication follows directly from the fact that $L_{x_\infty}\mu(x_\infty) = 0$ by Theorem 3.3. This proves Theorem 13.1. □

In his PhD thesis [72] Gábor Székelyhidi found a criterion for the existence of a critical point of the moment map squared in the complexified group orbit in terms of polystability with respect to the action of a suitable subgroup. To describe his criterion we recall the notations

$$\mathrm{G}_x := \{u \in \mathrm{G} \mid ux = x\}, \qquad \mathrm{G}^c_x := \left\{g \in \mathrm{G}^c \mid gx = x\right\}$$

for the compact and complex stabilizer subgroups. Recall also that G^c_x is the complexification of G_x whenever $\mu(x) = 0$ (Lemma 2.3). However, in general, the *complexified* stabilizer subgroup $(\mathrm{G}_x)^c$ is a proper subgroup of the *complex* stabilizer subgroup G^c_x. The latter may not even be reductive.

Throughout a *torus* is a compact connected abelian Lie group. The Székelyhidi criterion requires the choice of a maximal torus

$$\mathrm{T} \subset \mathrm{G}^c_x$$

and it may not be possible to choose this torus such that it is contained in G. However, if $\mathrm{T} \subset \mathrm{G}^c$ is any torus and $\mathfrak{t} := \mathrm{Lie}(\mathrm{T})$ is its Lie algebra, then $\mathfrak{t} \setminus \{0\} \subset \mathscr{T}^c$ and so the G^c-invariant pairing

$$\langle \zeta_1, \zeta_2 \rangle_c := \langle \mathrm{Re}(\zeta_1), \mathrm{Re}(\zeta_2) \rangle - \langle \mathrm{Im}(\zeta_1), \mathrm{Im}(\zeta_2) \rangle \tag{13.1}$$

on $\mathfrak{g}^c$ is positive definite on $\mathfrak{t}$ by Lemma 5.7. This implies that there exists a unique linear projection $\Pi_\mathrm{T} : \mathfrak{g}^c \to \mathfrak{t}$ such that

$$\langle \zeta - \Pi_\mathrm{T}(\zeta), \tau \rangle_c = 0 \qquad \text{for all } \zeta \in \mathfrak{g}^c \text{ and all } \tau \in \mathfrak{t}. \tag{13.2}$$

With this understood, we introduce the following notation.

Let $x \in X$ and let $\mathrm{T} \subset \mathrm{G}^c_x$ be a torus with the Lie algebra $\mathfrak{t} := \mathrm{Lie}(\mathrm{T}) \subset \mathfrak{g}^c$. Let $\mathrm{G}^c_T \subset \mathrm{G}^c$ be the identity component of the centralizer of T (the subgroup of all elements of G^c that commute with each element of T), i.e.

$$\begin{aligned} \mathfrak{g}^c_\mathrm{T} &:= \left\{\zeta \in \mathfrak{g}^c \,\middle|\, [\zeta, \tau] = 0 \text{ for all } \tau \in \mathfrak{t}\right\}, \\ \mathrm{G}^c_\mathrm{T} &:= \left\{ g(1) \,\middle|\, \begin{array}{l} g : [0,1] \to \mathrm{G} \text{ is a smooth path} \\ \text{such that } g(0) = \mathbb{1} \text{ and} \\ \dot g(t) g(t)^{-1} \in \mathfrak{g}^c_\mathrm{T} \text{ for all } t \in [0,1] \end{array} \right\}. \end{aligned} \tag{13.3}$$

By the Closed Subgroup Theorem G^c_T is a Lie subgroup of G^c with the Lie algebra $\mathrm{Lie}(\mathrm{G}^c_\mathrm{T}) = \mathfrak{g}^c_\mathrm{T}$. For $x \in X$ and $\zeta \in \mathscr{T}^c \cap \mathfrak{g}^c_\mathrm{T} \setminus \mathfrak{t}$ the (μ, T)-**weight** of the pair (x, ζ) is defined by

$$w_{\mu,\mathrm{T}}(x, \zeta) := \lim_{t\to\infty} \langle \mu(\exp(\mathbf{i}t\zeta)x), \mathrm{Re}(\zeta - \Pi_\mathrm{T}(\zeta))\rangle. \tag{13.4}$$

With this terminology in place we are in a position to formulate the main results of this section.

Theorem 13.2 (Generalized Székelyhidi Criterion) *Let $x \in X$, let $\mathrm{T} \subset \mathrm{G}^c_x$ be a maximal torus with the Lie algebra $\mathfrak{t} := \mathrm{Lie}(\mathrm{T})$, and let $g \in \mathrm{G}^c$ such that $g\mathrm{T}g^{-1} \subset \mathrm{G}$. Then the following are equivalent.*

- **(i)** *$\mathrm{G}^c(x)$ contains a critical point of the square of the moment map.*
- **(ii)** *There exists an element $h \in \mathrm{G}^c_\mathrm{T}$ such that $g^{-1}\mu(ghx)g \in \mathfrak{t}$.*

Proof Assume that $h \in \mathrm{G}^c_\mathrm{T}$ satisfies $g^{-1}\mu(ghx)g \in \mathfrak{t} \subset \ker L^c_x$. Then $h\tau h^{-1} = \tau$ and hence $L^c_{hx}\tau = hL^c_x\tau = 0$ for all $\tau \in \mathfrak{t}$. Thus $L_{ghx}\mu(ghx) = gL^c_{hx}(g^{-1}\mu(ghx)g) = 0$. This shows that (ii) implies (i). The converse is proved on page 120. □

Theorem 13.3 (Székelyhidi Moment-Weight Inequality) *Let $x \in X$, let $\mathrm{T} \subset \mathrm{G}^c_x$ be a torus with the Lie algebra $\mathfrak{t} := \mathrm{Lie}(\mathrm{T})$, and let $g \in \mathrm{G}^c$ such that $g\mathrm{T}g^{-1} \subset \mathrm{G}$. Then*

$$\inf_{h\in\mathrm{G}^c} |\mu(hx)| \geq |\Pi_\mathrm{T}(g^{-1}\mu(gx)g)|_c$$

and

$$\sup_{\zeta\in\mathscr{T}^c\cap\mathfrak{g}^c_\mathrm{T}\setminus\mathfrak{t}} \frac{-w_{\mu,\mathrm{T}}(x,\zeta)}{\sqrt{|\zeta|_c^2 - |\Pi_\mathrm{T}(\zeta)|_c^2}} \leq \inf_{h\in\mathrm{G}^c} \sqrt{|\mu(hx)|^2 - |\Pi_\mathrm{T}(g^{-1}\mu(gx)g)|_c^2}. \tag{13.5}$$

Moreover, the supremum on the left is attained, it remains unchanged when taken over all $\zeta \in \mathscr{T}^c \cap \mathfrak{g}^c_\mathrm{T} \setminus \mathfrak{t}$ that satisfy $\exp(\zeta) \in \mathrm{T}$, $g\zeta g^{-1} \in \mathfrak{g}$, $\Pi_\mathrm{T}(\zeta) = 0$, and equality holds in (13.5) if and only if the left hand side is nonnegative.

Proof See page 124. □

Theorem 13.4 (Hilbert–Mumford Criterion: Critical Orbits) *Let $x \in X$ and fix a maximal torus $\mathrm{T} \subset \mathrm{G}^c_x$ with Lie the algebra $\mathfrak{t} := \mathrm{Lie}(\mathrm{T})$. Choose an element $g \in \mathrm{G}^c$ such that $g\mathrm{T}g^{-1} \subset \mathrm{G}$. Then the following are equivalent.*

- **(i)** *$\mathrm{G}^c(x)$ contains a critical point of the square of the moment map.*
- **(ii)** *If $\zeta \in \mathscr{T}^c$ satisfies*

$$[\zeta, \mathfrak{t}] = 0, \qquad \exp(\zeta) \in \mathrm{T}, \qquad g\zeta g^{-1} \in \mathfrak{g}, \qquad \Pi_\mathrm{T}(\zeta) = 0,$$

then $w_\mu(x, \zeta) \geq 0$ *and*

$$w_\mu(x, \zeta) = 0 \qquad \Longrightarrow \qquad \lim_{t\to\infty} \exp(\mathbf{i}t\zeta)x \in \mathrm{G}^c_\mathrm{T}(x).$$

(iii) *If* $\zeta \in \mathscr{T}^c \setminus \mathfrak{t}$ *satisfies* $[\zeta, \mathfrak{t}] = 0$, *then* $w_{\mu,\mathrm{T}}(x, \zeta) \geq 0$ *and*

$$w_{\mu,\mathrm{T}}(x, \zeta) = 0 \qquad \Longrightarrow \qquad \lim_{t\to\infty} \exp(\mathbf{i}t\zeta)x \in \mathrm{G}^c_\mathrm{T}(x).$$

Proof That (iii) implies (ii) follows from the fact that

$$w_{\mu,\mathrm{T}}(x, \zeta) = w_\mu(x, \zeta)$$

whenever $\zeta \in \mathscr{T}^c$ satisfies $[\zeta, \mathfrak{t}] = 0$ and $\Pi_\mathrm{T}(\zeta) = 0$ (see Eq. (13.4)). That (ii) implies (i) and that (i) implies (iii) will be proved on page 126. □

We will see below that the Székelyhidi criterion in Theorem 13.2 is a restatement of the polystability condition on x with respect to the action of a suitable quotient group on a suitable submanifold of X. In the case

$$g = \mathbb{1},$$

where T is a subgroup of G, the Lie group in question is the quotient

$$\mathrm{G}_\mathrm{T}/\mathrm{T}$$

of the identity component $\mathrm{G}_\mathrm{T} \subset \mathrm{G}$ of the centralizer of T by the torus T, and the submanifold X_T, on which it acts, consists of all elements of X that contain the torus T in their stabilizer subgroup. Then the moment map descends to an equivariant moment map

$$\mu_\mathrm{T} : X_\mathrm{T} \to \mathfrak{g}_\mathrm{T}/\mathfrak{t}$$

for this action and, in the case $g = \mathbb{1}$, condition (ii) in Theorem 13.2 says that $x \in X_\mathrm{T}$ is μ_T-polystable (see Lemma 13.9). Theorem 13.3 states the corresponding moment-weight inequality for μ_T and Theorem 13.4 adapts the Hilbert–Mumford criterion for polystability to this setting. The proofs of all three theorems reduce to the case $g = \mathbb{1}$, either by replacing (x, T) with

$$(gx, g\mathrm{T}g^{-1})$$

or, equivalently, by leaving the pair (x, T) unchanged while replacing $(\mathrm{G}, \omega, \mu)$ with the triple $(g^*\mathrm{G}, g^*\omega, g^*\mu)$, where

$$g^*\mathrm{G} := g^{-1}\mathrm{G}g, \qquad (g^*\omega)_x(\widehat{x}_1, \widehat{x}_2) := \omega_{gx}(g\widehat{x}_1, g\widehat{x}_2),$$

and

$$(g^*\mu)(x) := g^{-1}\mu(gx)g.$$

This is the first ingredient in the proof of the Székelyhidi criterion and is explained in Lemma 13.5 below. The second ingredient in the proof is the study of the action of the quotient group $\mathrm{G_T/T}$ on the submanifold X_T and its moment map $\mu_\mathrm{T} : X_\mathrm{T} \to \mathfrak{g}_\mathrm{T}/\mathfrak{t}$ in the case $\mathrm{T} \subset \mathrm{G}$. This is the content of the Lemma 13.9. The third ingredient, required for the proof of the generalized Székelyhidi moment-weight inequality and of the Hilbert–Mumford criterion for critical orbits, is the study of the toral generators for the quotient group (Lemma 13.11) and the corresponding Mumford numerical invariants. These are shown in Lemma 13.12 to be the (μ, T)-weights in (13.4). After these preparations we are ready to prove the main theorems of this section. The proof for $g = \mathbb{1}$ will again use the gradient flow of the moment map squared and the Kempf–Ness function as the central technical ingredients.

Conjugation and the Balancing Condition

The purpose of the present subsection is to explain why it suffices to prove Theorems 13.2, 13.3, and 13.4 in the case

$$g = \mathbb{1}.$$

Lemma 13.5 *Fix an element $g \in \mathrm{G}^c$, define*

$$\widetilde{\mathrm{G}} := g^{-1}\mathrm{G}g, \qquad \widetilde{\mathfrak{g}} := g^{-1}\mathfrak{g}g, \tag{13.6}$$

and define the map $\widetilde{\mu} : X \to \widetilde{\mathfrak{g}}$ by

$$\widetilde{\mu}(x) := g^{-1}\mu(gx)g \qquad \text{for } x \in X. \tag{13.7}$$

Let

$$\widetilde{\omega} := g^*\omega$$

be the pullback of ω under the diffeomorphism induced by g. Then the following holds.

(i) *The bilinear form* (13.1) *defines an invariant inner product on $\widetilde{\mathfrak{g}}$.*
(ii) *The 2-form $\widetilde{\omega}$ is a Kähler form on the complex manifold (X, J).*
(iii) *The Lie group $\widetilde{\mathrm{G}}$ acts on $(X, \widetilde{\omega}, J)$ by Kähler isometries.*
(iv) *The map $\widetilde{\mu} : X \to \widetilde{\mathfrak{g}}$ is $\widetilde{\mathrm{G}}$-equivariant.*
(v) *The map $\widetilde{\mu} : X \to \widetilde{\mathfrak{g}}$ is a moment map for the $\widetilde{\mathrm{G}}$-action on $(X, \widetilde{\omega})$.*
(vi) *Let $x \in X$ and let $\mathrm{T} \subset \mathrm{G}^c_x$ be a torus with Lie algebra $\mathfrak{t}$. Then*

$$w_{\widetilde{\mu},\mathrm{T}}(x, \zeta) = w_{\mu,\mathrm{T}}(x, \zeta) \tag{13.8}$$

for every $\zeta \in \mathscr{T}^c \cap \mathfrak{g}^c_\mathrm{T} \setminus \mathfrak{t}$.

Proof We prove part (i). The bilinear form (13.1) on $\mathfrak{g}^c$ is symmetric by definition and, since $\widetilde{\mathfrak{g}} \setminus \{0\} \subset \mathscr{T}^c$, its restriction to $\widetilde{\mathfrak{g}}$ is positive definite by Lemma 5.7. This proves (i).

Part (ii) follows from the fact that ω is a Kähler form on (X, J) and that the diffeomorphism induced by g preserves the complex structure.

We prove part (iii). Fix an element $\widetilde{u} \in \widetilde{\mathrm{G}}$ and define

$$u := g\widetilde{u}g^{-1}.$$

Then $u \in \mathrm{G}$ by (13.6) and hence

$$\widetilde{u}^*\widetilde{\omega} = \widetilde{u}^* g^* \omega = (g\widetilde{u})^*\omega = (ug)^*\omega = g^* u^* \omega = g^* \omega = \widetilde{\omega}.$$

Moreover, the $\widetilde{\mathrm{G}}$-action preserves the complex structure because $\widetilde{\mathrm{G}} \subset \mathrm{G}^c$. This proves (iii).

We prove part (iv). Let $x \in X$ and $\widetilde{u} \in \widetilde{\mathrm{G}}$, and define $u := g\widetilde{u}g^{-1} \in \mathrm{G}$ as above. Then, by (13.7),

$$\begin{aligned}
\widetilde{\mu}(\widetilde{u}x) &= g^{-1}\mu(g\widetilde{u}x)g \\
&= g^{-1}\mu(ugx)g \\
&= g^{-1}u\mu(gx)u^{-1}g \\
&= \widetilde{u}g^{-1}\mu(gx)g\widetilde{u}^{-1} \\
&= \widetilde{u}\widetilde{\mu}(x)\widetilde{u}^{-1}
\end{aligned}$$

and this proves (iv).

We prove part (v). Fix elements $x \in X$, $\widehat{x} \in T_x X$, and $\widetilde{\xi} \in \widetilde{\mathfrak{g}}$, and define

$$\xi := g\widetilde{\xi}g^{-1}.$$

Then $\xi \in \mathfrak{g}$ by (13.6) and

$$\begin{aligned}
\widetilde{\omega}_x(L_x^c\widetilde{\xi}, \widehat{x}) &= \omega_{gx}(gL_x^c\widetilde{\xi}, g\widehat{x}) \\
&= \omega_{gx}(L_{gx}(g\widetilde{\xi}g^{-1}), g\widehat{x}) \\
&= \omega_{gx}(L_{gx}\xi, g\widehat{x}) \\
&= \langle d\mu(gx)g\widehat{x}, \xi\rangle \\
&= \langle g^{-1}(d\mu(gx)g\widehat{x})g, g^{-1}\xi g\rangle_c \\
&= \langle d\widetilde{\mu}(x)\widehat{x}, \widetilde{\xi}\rangle_c.
\end{aligned}$$

Here the penultimate equality follows from Lemma 5.7 and the last equality follows from (13.7). This proves (v).

We prove part (vi). Let $x \in X$, let $\mathrm{T} \subset \mathrm{G}_x^c$ be a torus with Lie algebra $\mathfrak{t}$, let $\zeta \in \mathscr{T}^c \setminus \mathfrak{t}$ with $[\zeta, \mathfrak{t}] = 0$, and define $x^+ := \lim_{t\to\infty} \exp(\mathbf{i}t\zeta)x$. Then

$$x^+ = \lim_{t\to\infty} \exp(\mathbf{i}t(\zeta - \Pi_{\mathrm{T}}(\zeta)))x$$

because $L_x^c \Pi_{\mathrm{T}}(\zeta) = 0$ and ζ commutes with $\Pi_{\mathrm{T}}(\zeta)$. Hence $L_{x^+}^c(\zeta - \Pi_{\mathrm{T}}(\zeta)) = 0$ by Lemma 5.4, and so

$$\begin{aligned} w_{\mu,\mathrm{T}}(x,\zeta) &= \langle \mu(x^+), \zeta - \Pi_{\mathrm{T}}(\zeta)\rangle_c \\ &= \langle \mu(gx^+), g(\zeta - \Pi_{\mathrm{T}}(\zeta))g^{-1}\rangle_c \\ &= \langle g^{-1}\mu(gx^+)g, \zeta - \Pi_{\mathrm{T}}(\zeta)\rangle_c \\ &= w_{\widetilde{\mu},\mathrm{T}}(x,\zeta). \end{aligned}$$

Here the second equality follows from Lemma 5.8, the third equality follows from Lemma 5.7, and the first and last equalities follow from the definition of the relative weights in (13.4). This proves (vi) and Lemma 13.5. □

Definition 13.6 An element $x \in X$ is called μ-**balanced** iff there exists a maximal torus $\mathrm{T} \subset \mathrm{G}_x^c$ such that $\mathrm{T} \subset \mathrm{G}$.

Here we slightly abuse notation because the balancing condition depends only on the maximal compact subgroup of G^c but not on the choice of the moment map. Lemma 13.5 shows that it suffices to prove Theorems 13.2, 13.3, and 13.4 under the assumption that x is μ-balanced and $g = \mathbb{1}$.

Example 13.7

(i) If $g \in \mathrm{G}^c$ belongs to the center of G^c, then the maximal compact subgroup $\widetilde{\mathrm{G}} = g^{-1}\mathrm{G}g$ in Lemma 13.5 is equal to G, however, the resulting symplectic form $\widetilde{\omega} := g^*\omega$ and the moment map $\widetilde{\mu} = g^*\mu$ may well be different from ω and μ, respectively.

(ii) Every fixed point of the G^c-action is μ-balanced and so is every point with discrete isotropy in G^c.

(iii) If G is abelian then every torus in G^c is necessarily contained in G and so every element of X is μ-balanced.

(iv) Every G^c-orbit contains a μ-balanced element. Let $x \in X$ and $\mathrm{T} \subset \mathrm{G}_x^c$ be a maximal torus. By Lemma C.3 there is a $g \in \mathrm{G}^c$ such that $g\mathrm{T}g^{-1} \subset \mathrm{G}$. Then $g\mathrm{T}g^{-1}$ is a maximal torus in $g\mathrm{G}_x^c g^{-1} = \mathrm{G}_{gx}^c$ and hence gx is μ-balanced.

(v) Consider the diagonal action of $\mathrm{G} = \mathrm{SO}(3)$ on $X = (S^2)^n$. Then $\mathrm{G}^c = \mathrm{PSL}(2,\mathbb{C})$ is the group of Möbius transformations, the moment map $\mu : (S^2)^n \to \mathbb{R}^3$ is given by $\mu(x) = \sum_{i=1}^n x_i$ for $x = (x_1, \dots, x_n) \in (S^2)^n$, and its critical points are the n-tuples $x \in (S^2)^n$ that satisfy $x_i = \pm x_j$ for all i and j. Now let $x \in (S^2)^n$ such that $x_2 = \cdots = x_n \neq \pm x_1$. Then $\mathrm{G}_x = \{\mathbb{1}\}$ and $\mathrm{G}_x^c \cong \mathbb{C}^*$. Thus G_x^c contains a unique maximal torus $\mathrm{T} \not\subset \mathrm{G}$ and so x is not μ-balanced. In fact, an element of $(S^2)^n$ with nontrivial stabilizer subgroup in G^c is μ-balanced if and only if it is a critical point of the square of the moment map.

Corollary 13.8 (Generalized Székelyhidi Criterion) *Let $x \in X$ be a μ-balanced element and let $\mathrm{T} \subset \mathrm{G}_x$ be a maximal torus with the Lie algebra $\mathfrak{t} := \mathrm{Lie}(\mathrm{T})$. Then*

$$\inf_{g \in \mathrm{G}^c} |\mu(gx)| \geq |\Pi_{\mathrm{T}}(\mu(x))|$$

and

$$\sup_{\xi \in \mathfrak{g} \setminus \{0\},\, [\xi, \mathfrak{t}] = 0,\, \xi \perp \mathfrak{t}} \frac{-w_\mu(x, \xi)}{|\xi|} \leq \inf_{g \in \mathrm{G}^c} \sqrt{|\mu(gx)|^2 - |\Pi_{\mathrm{T}}(\mu(x))|^2}. \tag{13.9}$$

Moreover, the following are equivalent.

(i) *$\mathrm{G}^c(x)$ contains a critical point of the square of the moment map.*
(ii) *There exists an element $h \in \mathrm{G}_{\mathrm{T}}^c$ such that $\mu(hx) \in \mathfrak{t}$.*
(iii) *If $\xi \in \mathfrak{g} \setminus \{0\}$ satisfies $[\xi, \mathfrak{t}] = 0$ and $\xi \perp \mathfrak{t}$ then $w_\mu(x, \xi) \geq 0$ and*

$$w_\mu(x, \xi) = 0 \quad \Longrightarrow \quad \lim_{t \to \infty} \exp(\mathbf{i} t \xi) x \in \mathrm{G}_{\mathrm{T}}^c(x).$$

Proof Theorems 13.2, 13.3, and 13.4. □

Centralizer and Quotient Group

For a torus $\mathrm{T} \subset \mathrm{G}$ with the Lie algebra $\mathfrak{t} := \mathrm{Lie}(\mathrm{T}) \subset \mathfrak{g}$ denote by $\mathrm{G}_T \subset \mathrm{G}$ the identity component of the centralizer of T (the subgroup of all elements of G that commute with each element of T), i.e.

$$\mathfrak{g}_{\mathrm{T}} := \left\{ \xi \in \mathfrak{g} \,\middle|\, [\xi, \tau] = 0 \text{ for all } \tau \in \mathfrak{t} \right\},$$

$$\mathrm{G}_{\mathrm{T}} := \left\{ u(1) \,\middle|\, \begin{array}{l} u : [0,1] \to \mathrm{G} \text{ is a smooth path} \\ \text{such that } u(0) = \mathbb{1} \text{ and} \\ \dot{u}(t)u(t)^{-1} \in \mathfrak{g}_{\mathrm{T}} \text{ for all } t \in [0,1] \end{array} \right\}. \tag{13.10}$$

By the Closed Subgroup Theorem, G_{T} is a Lie subgroup of G with the Lie algebra $\mathrm{Lie}(\mathrm{G}_{\mathrm{T}}) = \mathfrak{g}_{\mathrm{T}}$. Moreover, T is a subgroup of the center of G_{T}. Thus T is a normal subgroup of G_{T} and $\mathrm{G}_{\mathrm{T}}/\mathrm{T}$ is a compact Lie group with the Lie algebra $\mathrm{Lie}(\mathrm{G}_{\mathrm{T}}/\mathrm{T}) = \mathfrak{g}_{\mathrm{T}}/\mathfrak{t}$. For $\xi \in \mathfrak{g}_{\mathrm{T}}$ denote by

$$[\xi]_{\mathrm{T}} := \xi + \mathfrak{t} \in \mathfrak{g}_{\mathrm{T}}/\mathfrak{t}$$

the equivalence class of ξ. The quotient group acts on the space

$$X_{\mathrm{T}} := \left\{ x \in X \,\middle|\, \mathrm{T} \subset \mathrm{G}_x \right\} \tag{13.11}$$

of all elements of X that contain the torus T in their stabilizer subgroup. The next lemma shows that X_{T} is a complex submanifold of X and that the $\mathrm{G}_{\mathrm{T}}/\mathrm{T}$-action on X_{T} is Hamiltonian.

Lemma 13.9 *Let* $\mathrm{T} \subset \mathrm{G}$ *be a torus with the Lie algebra* $\mathfrak{t}$, *let* $\mathrm{G}_{\mathrm{T}}, \mathfrak{g}_{\mathrm{T}}$ *be as in* (13.10), *and let* $\mathrm{G}^c_{\mathrm{T}}$, $\mathfrak{g}^c_{\mathrm{T}}$ *be as in* (13.3). *Then the following holds.*

(i) *The formula*

$$\langle [\xi]_{\mathrm{T}}, [\eta]_{\mathrm{T}} \rangle_{\mathfrak{g}_{\mathrm{T}}/\mathfrak{t}} := \langle \xi, \eta - \Pi_{\mathrm{T}}(\eta) \rangle \tag{13.12}$$

for $\xi, \eta \in \mathfrak{g}_{\mathrm{T}}$ *defines an invariant inner product on* $\mathfrak{g}_{\mathrm{T}}/\mathfrak{t}$.

(ii) *If* $\zeta \in \mathscr{T}^c \cap \mathfrak{g}^c_{\mathrm{T}} \setminus \mathfrak{t}$ *then* $\zeta + \tau \in \mathscr{T}^c \cap \mathfrak{g}^c_{\mathrm{T}} \setminus \mathfrak{t}$ *for all* $\tau \in \mathfrak{t}$ *and*

$$\mathrm{Im}(\zeta) \perp \mathfrak{t}.$$

(iii) *The set* X_{T} *in* (13.11) *is a closed* G_{T}*-invariant complex submanifold of* X *and* $\mu(X_{\mathrm{T}}) \subset \mathfrak{g}_{\mathrm{T}}$.

(iv) *The quotient group* $\mathrm{G}_{\mathrm{T}}/\mathrm{T}$ *acts on* X_{T} *by Kähler isometries, and the action is generated by the* $\mathrm{G}_{\mathrm{T}}/\mathrm{T}$*-equivariant moment map*

$$\mu_{\mathrm{T}} : X_{\mathrm{T}} \to \mathfrak{g}_{\mathrm{T}}/\mathfrak{t}, \qquad \mu_{\mathrm{T}}(x) := [\mu(x)]_{\mathrm{T}} = \mu(x) + \mathfrak{t}. \tag{13.13}$$

Proof See page 114. □

In preparation for the proof of Lemma 13.9 we establish some basic properties of the projections Π_{T}.

Lemma 13.10 *Let* $\mathrm{T} \subset \mathrm{G}^c$ *be a torus and let* $\Pi_{\mathrm{T}} : \mathfrak{g}^c \to \mathfrak{t}$ *be the projection defined by* (13.2). *Fix two elements* $g \in \mathrm{G}^c$ *and* $\zeta \in \mathfrak{g}^c$. *Then*

$$\Pi_{g\mathrm{T}g^{-1}}(g\zeta g^{-1}) = g\Pi_{\mathrm{T}}(\zeta)g^{-1}. \tag{13.14}$$

Moreover, if $x \in X$ *satisfies* $\mathrm{T} \subset \mathrm{G}^c_x$ *then*

$$\langle \mu(gx), \mathrm{Re}(\Pi_{g\mathrm{T}g^{-1}}(g\zeta g^{-1})) \rangle = \langle \mu(x), \mathrm{Re}(\Pi_{\mathrm{T}}(\zeta)) \rangle \tag{13.15}$$

Proof Let $\mathfrak{t} := \mathrm{Lie}(\mathrm{T}) \subset \mathfrak{g}^c$ and define $\tau := \Pi_{\mathrm{T}}(\zeta) \in \mathfrak{t}$. Then $\langle \zeta - \tau, \mathfrak{t} \rangle_c = 0$ by (13.2) and hence, by Lemma 5.8,

$$\langle g\zeta g^{-1} - g\tau g^{-1}, g\mathfrak{t}g^{-1} \rangle_c = 0.$$

Thus $g\tau g^{-1} = \Pi_{g\mathrm{T}g^{-1}}(g\zeta g^{-1})$ and this proves (13.14).

Now let $x \in X$ such that $\mathrm{T} \subset \mathrm{G}^c_x$. Then $L^c_x \tau = 0$ and hence it follows from Lemma 5.8 that $\langle \mu(x), \mathrm{Re}(\tau) \rangle = \langle \mu(gx), \mathrm{Re}(g\tau g^{-1}) \rangle$. Thus Eq. (13.15) follows from (13.14) and this proves Lemma 13.10. □

Proof of Lemma 13.9 We prove part (i). Let $\xi, \eta \in \mathfrak{g}_{\mathrm{T}}$ and $u \in \mathrm{G}_{\mathrm{T}}$. Then

$$\Pi_{\mathrm{T}}(u\eta u^{-1}) = \Pi_{\mathrm{T}}(\eta)$$

by Lemma 13.10 and so

$$\begin{aligned}\langle [u\xi u^{-1}]_{\mathrm{T}}, [u\eta u^{-1}]_{\mathrm{T}}\rangle_{\mathfrak{g}_{\mathrm{T}}/\mathfrak{t}} &= \langle u\xi u^{-1}, u\eta u^{-1} - \Pi_{\mathrm{T}}(\eta)\rangle \\ &= \langle \xi, \eta - \Pi_{\mathrm{T}}(\eta)\rangle \\ &= \langle [\xi]_{\mathrm{T}}, [\eta]_{\mathrm{T}}\rangle_{\mathfrak{g}_{\mathrm{T}}/\mathfrak{t}}.\end{aligned}$$

This proves (i).

We prove part (ii). Fix an element $\zeta \in \mathscr{T}^c \cap \mathfrak{g}_{\mathrm{T}}^c \setminus \mathfrak{t}$ and let $\tau \in \mathfrak{t}$. Then ζ and τ are two commuting toral generators and $\zeta + \tau \neq 0$. Hence $\zeta + \tau$ is again a toral generator, commutes with $\mathfrak{t}$, and is not an element of $\mathfrak{t}$. Thus $\zeta + \tau \in \mathscr{T}^c \cap \mathfrak{g}_{\mathrm{T}}^c \setminus \mathfrak{t}$. This implies $\langle \mathrm{Re}(\zeta) + \tau, \mathrm{Im}(\zeta)\rangle = 0$ for all $\tau \in \mathfrak{t}$ by Lemma 5.7. Hence $\mathrm{Im}(\zeta) \perp \mathfrak{t}$ and this proves (ii).

We prove part (iii). That X_{T} is a closed subset of X follows directly from the definitions, and that it is a submanifold of X with the tangent spaces

$$T_x X_{\mathrm{T}} = \left\{\widehat{x} \in T_x X \,\middle|\, a\widehat{x} = \widehat{x} \text{ for all } a \in \mathrm{T}\right\}$$

for $x \in X_{\mathrm{T}}$ is a general fact about smooth group actions. That X_{T} is a complex submanifold follows from the fact that the T-action preserves the complex structure. If $x \in X_{\mathrm{T}}$ and $g \in \mathrm{G}_{\mathrm{T}}$, then $ax = x$ for all $a \in \mathrm{T}$, hence $agx = gax = gx$ for all $a \in \mathrm{T}$ because a and g commute, and hence $gx \in X_{\mathrm{T}}$. Thus X_{T} is G_{T}-invariant. Moreover, the group T acts trivially on X_{T} by definition, and hence the action of G_{T} on X_{T} by Kähler isometries descends to an action of the quotient group $\mathrm{G}_{\mathrm{T}}/\mathrm{T}$.

If $x \in X_{\mathrm{T}}$, then $\mathfrak{t} \subset \ker L_x$, hence

$$[\mu(x), \tau] = -d\mu(x)L_x\tau = 0$$

for all $\tau \in \mathfrak{t}$ by (2.6), and so $\mu(x) \in \mathfrak{g}_{\mathrm{T}}$. This shows that

$$\mu(X_{\mathrm{T}}) \subset \mathfrak{g}_{\mathrm{T}}$$

and so the map $\mu_{\mathrm{T}} : X_{\mathrm{T}} \to \mathfrak{g}_{\mathrm{T}}/\mathfrak{t}$ in (13.13) is well defined. That it is $\mathrm{G}_{\mathrm{T}}/\mathrm{T}$-equivariant follows directly from the G-equivariance of map $\mu : X \to \mathfrak{g}$. This proves (iii).

We prove part (iv). Fix elements $x \in X_{\mathrm{T}}$, $\widehat{x} \in T_x X_{\mathrm{T}}$, and $\xi \in \mathfrak{g}_{\mathrm{T}}$. Then

$$\begin{aligned}\langle d\mu_{\mathrm{T}}(x)\widehat{x}, [\xi]_{\mathrm{T}}\rangle_{\mathfrak{g}_{\mathrm{T}}/\mathfrak{t}} &= \langle d\mu(x)\widehat{x}, \xi - \Pi_{\mathrm{T}}(\xi)\rangle \\ &= \langle d\mu(x)\widehat{x}, \xi\rangle \\ &= \omega(L_x\xi, \widehat{x}).\end{aligned}$$

Here the second equality follows from the equation $d\mu(x)^* = JL_x$ in (2.6) and the fact that $\mathfrak{t} \subset \ker L_x$. This proves (iv) and Lemma 13.9. □

By definition, an element $x \in X_{\mathrm{T}}$ is μ_{T}-polystable if and only if there exists an element $h \in \mathrm{G}^c_{\mathrm{T}}$ such that $\mu(hx) \in \mathfrak{t}$. Thus the generalized Székelyhidi criterion in Theorem 13.2 asserts that, when $\mathrm{T} \subset \mathrm{G}^c_x$ is a maximal torus and $\mathrm{T} \subset \mathrm{G}$, the complexified group orbit $\mathrm{G}^c(x)$ contains a critical point of the square of the moment map if and only if $x \in X_{\mathrm{T}}$ is μ_{T}-polystable. If G is a torus, this follows directly definitions.

The (μ, T)-Weights

This subsection establishes the basic properties of the relative (μ, T)-weights, in preparation for the proof of the Székelyhidi moment-weight inequality and of the Hilbert–Mumford criterion for critical orbits. The first step is to characterize of the toral generators in the quotient Lie algebra $\mathfrak{g}^c_{\mathrm{T}}/\mathfrak{t}^c$.

Lemma 13.11 *Let* $\mathrm{T} \subset \mathrm{G}^c$ *be a torus and let* $\mathrm{G}^c_{\mathrm{T}}$ *and* $\mathfrak{g}^c_{\mathrm{T}}$ *be as in* (13.3). *Then* $\mathrm{G}^c_{\mathrm{T}}/\mathrm{T}^c$ *is a reductive Lie group with the Lie algebra*

$$\mathfrak{g}^c_{\mathrm{T}}/\mathfrak{t}^c = \mathrm{Lie}(\mathrm{G}^c_{\mathrm{T}}/\mathrm{T}^c).$$

Moreover, the set of toral generators in $\mathfrak{g}^c_{\mathrm{T}}/\mathfrak{t}^c$ *is given by*

$$\mathscr{T}^c_{\mathrm{G}^c_{\mathrm{T}}/\mathrm{T}^c} = \left\{[\zeta]_{\mathrm{T}^c} \,\middle|\, \zeta \in \mathscr{T}^c \cap \mathfrak{g}^c_{\mathrm{T}} \setminus \mathfrak{t}\right\}.$$

Here $[\zeta]_{\mathrm{T}^c} := \zeta + \mathfrak{t}^c$ *denotes the equivalence class of* $\zeta \in \mathfrak{g}^c_{\mathrm{T}}$ *in* $\mathfrak{g}^c_{\mathrm{T}}/\mathfrak{t}^c$.

Proof Assume first that $\mathrm{T} \subset \mathrm{G}$. Then $\mathrm{G}^c_{\mathrm{T}}/\mathrm{T}^c$ is a connected complex Lie group which contains $\mathrm{G}_{\mathrm{T}}/\mathrm{T}$ as a maximal compact subgroup, and whose Lie algebra is the complexification of $\mathfrak{g}_{\mathrm{T}}/\mathfrak{t} = \mathrm{Lie}(\mathrm{G}_{\mathrm{T}}/\mathrm{T})$. Thus $\mathrm{G}^c_{\mathrm{T}}/\mathrm{T}^c$ is the complexification of $\mathrm{G}_{\mathrm{T}}/\mathrm{T}$ and this proves (i).

Now let $\zeta \in \mathscr{T}^c \cap \mathfrak{g}^c_{\mathrm{T}} \setminus \mathfrak{t}$. Then the set

$$\mathrm{T}_\zeta := \overline{\{\exp(t\zeta) \,|\, t \in \mathbb{R}\}}$$

is a compact subgroup of $\mathrm{G}^c_{\mathrm{T}}$, and so projects to a compact subgroup of $\mathrm{G}^c_{\mathrm{T}}/\mathrm{T}^c$ which is generated by $[\zeta]_{\mathrm{T}^c}$. Since $\mathrm{Im}(\varPi_{\mathrm{T}}(\zeta)) \perp \mathfrak{t}$ by part (ii) of Lemma 13.9, we have $[\zeta]_{\mathrm{T}^c} \neq 0$ and so $[\zeta]_{\mathrm{T}^c}$ is a toral generator.

Conversely, choose an element $\zeta \in \mathfrak{g}^c_{\mathrm{T}}$ such that $[\zeta]_{\mathrm{T}^c}$ is a toral generator. Then $\zeta \notin \mathfrak{t}^c$ and we may assume without loss of generality that $\mathrm{Im}(\zeta) \perp \mathfrak{t}$. (If necessary, replace ζ by $\zeta - \mathbf{i}\mathrm{Im}(\varPi_{\mathrm{T}}(\zeta))$ without changing the equivalence class $[\zeta]_{\mathrm{T}^c}$.) Since $[\zeta]_{\mathrm{T}^c}$ is a toral generator, there exists an $h \in \mathrm{G}^c_{\mathrm{T}}$ such that $\mathrm{Im}(h\zeta h^{-1}) \in \mathfrak{t}$. Since $\mathrm{Im}(\zeta) \perp \mathfrak{t}$, it follows from Lemma 5.7 that $\mathrm{Im}(h\zeta h^{-1}) \perp \mathfrak{t}$ and therefore we have $\mathrm{Im}(h\zeta h^{-1}) = 0$. Hence $\zeta \in \mathscr{T}^c \cap \mathfrak{g}^c_{\mathrm{T}} \setminus \mathfrak{t}$ and this proves Lemma 13.11 under the assumption that $\mathrm{T} \subset \mathrm{G}$.

To prove the result in general, choose $g \in \mathrm{G}^c$ such that

$$\widetilde{\mathrm{T}} := g\mathrm{T}g^{-1} \subset \mathrm{G}.$$

Then $\mathrm{G}^c_{\mathrm{T}}/\mathrm{T}^c$ is isomorphic to $\mathrm{G}^c_{\widetilde{\mathrm{T}}}/\widetilde{\mathrm{T}}^c$ and hence is reductive. Now let $\zeta \in \mathfrak{g}^c_{\mathrm{T}}$. Then the equivalence class $[\zeta]_{\mathrm{T}^c} \in \mathfrak{g}^c_{\mathrm{T}}/\mathfrak{t}^c$ is a toral generator if and only if $[g\zeta g^{-1}]_{\widetilde{\mathrm{T}}^c}$ is a toral generator, and

$$\zeta \in \mathscr{T}^c \cap \mathfrak{g}^c_{\mathrm{T}} \setminus \mathfrak{t} \qquad \Longleftrightarrow \qquad g\zeta g^{-1} \in \mathscr{T}^c \cap \mathfrak{g}^c_{\widetilde{\mathrm{T}}} \setminus \widetilde{\mathfrak{t}}.$$

This proves Lemma 13.11. □

Lemma 13.12 *Let $x \in X$ and let $\mathrm{T} \subset \mathrm{G}^c_x$ be a torus with the Lie algebra $\mathfrak{t}$. Then the following holds.*

(i) *If $\zeta \in \mathscr{T}^c \cap \mathfrak{g}^c_{\mathrm{T}} \setminus \mathfrak{t}$, then*

$$w_{\mu,\mathrm{T}}(x, \zeta) = w_\mu(x, \zeta) - \langle \mu(x), \mathrm{Re}(\Pi_{\mathrm{T}}(\zeta))\rangle \tag{13.16}$$

(ii) *If $\zeta \in \mathscr{T}^c \cap \mathfrak{g}^c_{\mathrm{T}} \setminus \mathfrak{t}$ and $\tau \in \mathfrak{t}$, then $\zeta + \tau \in \mathscr{T}^c \cap \mathfrak{g}^c_{\mathrm{T}} \setminus \mathfrak{t}$ and*

$$w_{\mu,\mathrm{T}}(x, \zeta + \tau) = w_{\mu,\mathrm{T}}(x, \zeta) \tag{13.17}$$

(iii) *If $\zeta \in \mathscr{T}^c \cap \mathfrak{g}^c_{\mathrm{T}} \setminus \mathfrak{t}$ and $g \in \mathrm{G}^c$, then*

$$w_{\mu,\mathrm{T}}(x, \zeta) = w_{\mu,g\mathrm{T}g^{-1}}(gx, g\zeta g^{-1}) \tag{13.18}$$

(iv) *If $\mathrm{T} \subset \mathrm{G}_x$ and G_{T}, $\mathfrak{g}_{\mathrm{T}}$ are as in* (13.10) *and $\zeta \in \mathscr{T}^c \cap \mathfrak{g}^c_{\mathrm{T}} \setminus \mathfrak{t}$, then*

$$w_{\mu,\mathrm{T}}(x, \zeta) = w_{\mu_{\mathrm{T}}}(x, [\zeta]_{\mathrm{T}^c}). \tag{13.19}$$

Here the term on the right is the weight associated to the moment map (13.13) *for the action of $\mathrm{G}_{\mathrm{T}}/\mathrm{T}$ on the manifold X_{T} in Lemma 13.9.*

Proof We prove (i). Let $\zeta \in \mathscr{T}^c \cap \mathfrak{g}^c_{\mathrm{T}} \setminus \mathfrak{t}$. Then the group element

$$g := \exp(\mathbf{i}t\zeta) \in \mathrm{G}^c_{\mathrm{T}}$$

satisfies

$$g\zeta g^{-1} = \zeta, \qquad g\mathrm{T}g^{-1} = \mathrm{T}.$$

Hence it follows from Eq. (13.15) in Lemma 13.10 that

$$\langle \mu(\exp(\mathbf{i}t\zeta)x), \mathrm{Re}(\Pi_{\mathrm{T}}(\zeta))\rangle = \langle \mu(x), \mathrm{Re}(\Pi_{\mathrm{T}}(\zeta))\rangle$$

for all $t \in \mathbb{R}$. Thus, by (13.4), we have

$$\begin{aligned} w_{\mu,\mathrm{T}}(x,\zeta) &= \lim_{t\to\infty} \langle \mu(\exp(\mathbf{i}t\zeta)x), \mathrm{Re}(\zeta - \Pi_\mathrm{T}(\zeta))\rangle \\ &= \lim_{t\to\infty} \langle \mu(\exp(\mathbf{i}t\zeta)x), \mathrm{Re}(\zeta)\rangle - \langle \mu(x), \mathrm{Re}(\Pi_\mathrm{T}(\zeta))\rangle \\ &= w_\mu(x,\zeta) - \langle \mu(x), \mathrm{Re}(\Pi_\mathrm{T}(\zeta))\rangle. \end{aligned}$$

This proves (13.16) and (i).

We prove (ii). Let $\zeta \in \mathscr{T}^c \cap \mathfrak{g}_\mathrm{T}^c \setminus \mathfrak{t}$ and $\tau \in \mathfrak{t}$. Then

$$\zeta + \tau \in \mathscr{T}^c \cap \mathfrak{g}_\mathrm{T}^c \setminus \mathfrak{t}$$

by part (ii) of Lemma 13.9. Moreover, since $[\zeta,\tau] = 0$, we have

$$\exp(\mathbf{i}t(\zeta+\tau))x = \exp(\mathbf{i}t\zeta)\exp(\mathbf{i}t\tau)x = \exp(\mathbf{i}t\zeta)x$$

for all $t \in \mathbb{R}$, hence

$$x^+ := \lim_{t\to\infty} \exp(\mathbf{i}t\zeta)x = \lim_{t\to\infty} \exp(\mathbf{i}t(\zeta+\tau))x,$$

and so

$$\begin{aligned} w_{\mu,\mathrm{T}}(x,\zeta+\tau) &= \langle \mu(x^+), \mathrm{Re}(\zeta+\tau - \Pi_\mathrm{T}(\zeta+\tau))\rangle \\ &= \langle \mu(x^+), \mathrm{Re}(\zeta - \Pi_\mathrm{T}(\zeta))\rangle \\ &= w_{\mu,\mathrm{T}}(x,\zeta). \end{aligned}$$

by (13.4). This proves (13.17) and (ii).

We prove (iii). Let $\zeta \in \mathscr{T}^c \cap \mathfrak{g}_\mathrm{T}^c \setminus \mathfrak{t}$ and $g \in \mathrm{G}^c$. Then, by Lemma 13.10 and Theorem 5.3, we have

$$\begin{aligned} w_{\mu,\mathrm{T}}(x,\zeta) &= w_\mu(x,\zeta) - \langle \mu(x), \mathrm{Re}(\Pi_\mathrm{T}(\zeta))\rangle \\ &= w_\mu(gx, g\zeta g^{-1}) - \langle \mu(gx), \mathrm{Re}(\Pi_{g\mathrm{T}g^{-1}}(g\zeta g^{-1}))\rangle \\ &= w_{\mu,g\mathrm{T}g^{-1}}(gx, g\zeta g^{-1}). \end{aligned}$$

This proves (13.18) and (iii).

We prove (iv). Assume $\mathrm{T} \subset \mathrm{G}_x$ and let $\zeta \in \mathscr{T}^c \cap \mathfrak{g}_\mathrm{T}^c \setminus \mathfrak{t}$. Then it follows from part (ii) of Lemma 13.9 that

$$\Pi_\mathrm{T}(\mathrm{Im}(\zeta)) = 0.$$

Now define

$$x^+ := \lim_{t\to\infty} \exp(\mathbf{i}t\zeta)x.$$

Then, by (13.4), we have

$$\begin{aligned} w_{\mu,\mathrm{T}}(x,\zeta) &= \langle \mu(x^+), \mathrm{Re}(\zeta - \Pi_{\mathrm{T}}(\zeta))\rangle \\ &= \langle \mu(x^+), \mathrm{Re}(\zeta) - \Pi_{\mathrm{T}}(\mathrm{Re}(\zeta))\rangle \\ &= \langle \mu_{\mathrm{T}}(x^+), [\mathrm{Re}(\zeta)]_{\mathrm{T}}\rangle \\ &= w_{\mu_{\mathrm{T}}}(x, [\zeta]_{\mathrm{T}^c}). \end{aligned}$$

Here the second equality follows from the fact that $\mathfrak{t} \subset \mathfrak{g}$ and $\Pi_{\mathrm{T}}(\mathrm{Im}(\zeta)) = 0$, and the third equality follows from the definition of the inner product on $\mathfrak{g}_{\mathrm{T}}/\mathfrak{t}$ in (13.12) and the definition of the moment map $\mu_{\mathrm{T}} : X_{\mathrm{T}} \to \mathfrak{g}_{\mathrm{T}}/\mathfrak{t}$ in (13.13). This proves (13.19), part (iv), and Lemma 13.12. □

Proof of the Generalized Székelyhidi Criterion

The proof of Theorem 13.2 is based on the following three lemmas.

Lemma 13.13 *Let $x_0 \in X$, let $\mathrm{T} \subset \mathrm{G}_{x_0}$ be a torus with the Lie algebra $\mathfrak{t}$, and let G_{T} and $\mathfrak{g}_{\mathrm{T}}$ be as in* (13.10)*. Then the following holds for every element $g \in \mathrm{G}^c_{\mathrm{T}}$.*

(i) $\mathrm{T} \subset \mathrm{G}_{gx_0}$ *and* $\mathfrak{t} \subset \ker L_{gx_0}$.
(ii) $\mu(gx_0) \in \mathfrak{g}_{\mathrm{T}}$.
(iii) $\mu(gx_0) - \mu(x_0) \perp \mathfrak{t}$.

Proof Parts (i) and (ii) follow from Lemma 13.9 because $\mathrm{G}^c_{\mathrm{T}}(x_0) \subset X_{\mathrm{T}}$. To prove part (iii), choose a smooth path $g : [0,1] \to \mathrm{G}^c_{\mathrm{T}}$ with $g(0) = \mathbb{1}$ and define the paths $x : [0,1] \to X$ and $\eta, \xi : [0,1] \to \mathfrak{g}_{\mathrm{T}}$ by

$$x(t) := g(t)^{-1}x_0, \qquad \xi(t) + \mathbf{i}\eta(t) := g(t)^{-1}\dot{g}(t)$$

for $0 \le t \le 1$. Then

$$\dot{x} = -L_x\xi - JL_x\eta$$

and hence

$$\frac{d}{dt}\mu(x) = -d\mu(x)L_x\xi - d\mu(x)JL_x\eta = [\mu(x), \xi] - L_x^*L_x\eta.$$

Thus the inner product of $\frac{d}{dt}\mu(x)$ with any element $\tau \in \mathfrak{t}$ vanishes by part (i) and so the path $t \mapsto \langle \mu(x(t)), \tau\rangle$ is constant. This proves part (iii) and Lemma 13.13. □

Lemma 13.14 *Let $x_0 \in X$, let $x : \mathbb{R} \to X$ be the unique solution of* (3.2), *and define $x_\infty := \lim_{t\to\infty} x(t)$. Then the following holds for every $t \in \mathbb{R}$.*

(i) $\mathrm{G}_{x(t)} = \mathrm{G}_{x_0}$ *and* $\mathrm{G}_{x_0} \subset \mathrm{G}_{x_\infty}$.
(ii) $\ker L_{x(t)} = \ker L_{x_0}$ *and* $\ker L_{x_0} \subset \ker L_{x_\infty}$.
(iii) $[\mu(x(t)), \xi] = 0 = [\mu(x_\infty), \xi]$ *for all* $\xi \in \ker L_{x_0}$.
(iv) $\mu(x(t)) - \mu(x_0) \perp \ker L_{x_0}$ *and* $\mu(x_\infty) - \mu(x_0) \perp \ker L_{x_0}$.

Proof Let $u \in \mathrm{G}$ and $s \in \mathbb{R}$. Then the unique solution $y : \mathbb{R} \to X$ of the differential equation $\dot{y} = -JL_y\mu(y)$ with $y(s) = ux(s)$ is $y(t) = ux(t)$. This implies $\mathrm{G}_{x(s)} \subset \mathrm{G}_{x(t)}$ for all $s, t \in \mathbb{R}$. Interchange s and t to obtain

$$\mathrm{G}_{x(s)} = \mathrm{G}_{x(t)}$$

for all $s, t \in \mathbb{R}$. This proves (i) and (ii). To prove (iii), let $\xi \in \ker L_{x_0}$. Then

$$L_{x(t)}\xi = 0$$

for all $t \in \mathbb{R}$ by part (i), and hence

$$[\mu(x(t)), \xi] = -d\mu(x(t))L_{x(t)}\xi = 0$$

for all t by (2.6). This proves (iii). To prove (iv), we use (2.6) to compute

$$\frac{d}{dt}\mu(x(t)) = d\mu(x(t))\dot{x}(t) = -d\mu(x(t))JL_{x(t)}\mu(x(t)) = -L_{x(t)}^*L_{x(t)}\mu(x(t)).$$

Now fix an element $\xi \in \ker L_{x_0}$. Then $L_{x(t)}\xi = 0$ for all t by part (ii) and this implies

$$\frac{d}{dt}\langle\mu(x(t)), \xi\rangle = -\langle L_{x(t)}\xi, L_{x(t)}\mu(x(t))\rangle = 0.$$

This proves part (iv) and Lemma 13.14. □

Lemma 13.15 *Let $x_0 \in X$, let $x : \mathbb{R} \to X$ be the unique solution of* (3.2) *and define $x_\infty := \lim_{t\to\infty} x(t)$. Let $\mathrm{T} \subset \mathrm{G}_{x_0}$ be a torus with the Lie algebra $\mathfrak{t}$, and let G_T and $\mathfrak{g}_\mathrm{T}$ be as in* (13.10). *Then the following holds.*

(i) $\mu(x(t)) \in \mathfrak{g}_\mathrm{T}$ *for all* $t \in \mathbb{R}$ *and* $\mu(x_\infty) \in \mathfrak{g}_\mathrm{T}$.
(ii) $\mu(x(t)) - \mu(x_0) \perp \mathfrak{t}$ *for all* $t \in \mathbb{R}$ *and* $\mu(x_\infty) - \mu(x_0) \perp \mathfrak{t}$.
(iii) $x(t) \in \mathrm{G}_\mathrm{T}^c(x_0)$ *for all* $t \in \mathbb{R}$ *and* $x_\infty \in \overline{\mathrm{G}_\mathrm{T}^c(x_0)}$.

(iv) $|\mu(gx_0)| \geq |\mu(x_\infty)| \geq |\Pi_{\mathrm{T}}(\mu(x_0))|$ *for all* $g \in \mathrm{G}^c$ *and*

$$\begin{aligned} |\mu(x_\infty) - \Pi_{\mathrm{T}}(\mu(x_0))| &= \inf_{g \in \mathrm{G}^c_{\mathrm{T}}} \sqrt{|\mu(gx_0)|^2 - |\Pi_{\mathrm{T}}(\mu(x_0))|^2} \\ &= \inf_{g \in \mathrm{G}^c} \sqrt{|\mu(gx_0)|^2 - |\Pi_{\mathrm{T}}(\mu(x_0))|^2}. \end{aligned} \tag{13.20}$$

Proof Since $\mathfrak{t} \subset \ker L_{x_0}$, parts (i) and (ii) follow directly from parts (iii) and (iv) in Lemma 13.14. To prove part (iii) denote by $g : \mathbb{R} \to \mathrm{G}^c$ the unique solution of (3.3), so that

$$g^{-1}\dot{g} = \mathbf{i}\mu(x), \qquad g(0) = \mathbb{1}.$$

Since $\mu(x(t)) \in \mathfrak{g}_{\mathrm{T}}$ for all t by (i), this implies $g(t) \in \mathrm{G}^c_{\mathrm{T}}$, hence

$$x(t) = g(t)^{-1}x_0 \in \mathrm{G}^c_{\mathrm{T}}(x_0)$$

for all t by Lemma 3.2, and so

$$x_\infty = \lim_{t \to \infty} x(t) \in \overline{\mathrm{G}^c_{\mathrm{T}}(x_0)}.$$

This proves (iii). To prove part (iv) observe that

$$\Pi_{\mathrm{T}}(\mu(x_\infty)) = \Pi_{\mathrm{T}}(\mu(x_0))$$

by (ii) and hence, by Theorem 6.4,

$$|\Pi_{\mathrm{T}}(\mu(x_0))| = |\Pi_{\mathrm{T}}(\mu(x_\infty))| \leq |\mu(x_\infty)| = \inf_{g \in \mathrm{G}^c} |\mu(gx_0)|.$$

Since $x_\infty \in \overline{\mathrm{G}^c_{\mathrm{T}}(x_0)}$ by (iii), this implies

$$\begin{aligned} \sqrt{|\mu(x_\infty)|^2 - |\Pi_{\mathrm{T}}(\mu(x_0))|^2} &= \inf_{g \in \mathrm{G}^c} \sqrt{|\mu(gx_0)|^2 - |\Pi_{\mathrm{T}}(\mu(x_0))|^2} \\ &\leq \inf_{g \in \mathrm{G}^c_{\mathrm{T}}} \sqrt{|\mu(gx_0)|^2 - |\Pi_{\mathrm{T}}(\mu(x_0))|^2} \\ &\leq \sqrt{|\mu(x_\infty)|^2 - |\Pi_{\mathrm{T}}(\mu(x_0))|^2}. \end{aligned}$$

This proves (13.20) and Lemma 13.15. □

Proof of Theorem 13.2 (i) $\Longrightarrow$ (ii) Let $x_0 \in X$ such that $\mathrm{G}^c(x_0)$ contains a critical point of the square of the moment map and let $\mathrm{T} \subset \mathrm{G}^c_{x_0}$ be a maximal torus. By

Lemma 13.5 it suffices to assume that x_0 is μ-balanced and $\mathrm{T} \subset \mathrm{G}$. We must prove that there exists an element $h \in \mathrm{G}^c_\mathrm{T}$ such that

$$\mu(hx_0) \in \mathfrak{t} := \mathrm{Lie}(\mathrm{T}).$$

Let $x : \mathbb{R} \to X$ be the unique solution of (3.2) and define

$$x_\infty := \lim_{t\to\infty} x(t).$$

Then $L_{x_\infty}\mu(x_\infty) = 0$ by Theorem 3.3. We prove in five steps that

$$x_\infty \in \mathrm{G}^c_\mathrm{T}(x_0), \qquad \mu(x_\infty) \in \mathfrak{t}. \tag{13.21}$$

The μ-balanced condition is used in Step 2.

Step 1 *Let $g : \mathbb{R} \to \mathrm{G}^c$ be the unique solution of* (3.3)*. Then $g(t) \in \mathrm{G}^c_\mathrm{T}$ and we have $x(t) = g(t)^{-1}x_0$ for all $t \in \mathbb{R}$.*

Part (i) of Lemma 13.15 asserts that $\mu(x(t)) \in \mathfrak{g}_\mathrm{T}$ for all t. Hence it follows from (3.3) that $g(t)^{-1}\dot{g}(t) = \mathbf{i}\mu(x(t)) \in \mathfrak{g}^c_\mathrm{T}$ and this implies $g(t) \in \mathrm{G}^c_\mathrm{T}$ for all t. The formula $x(t) = g(t)^{-1}x_0$ follows from Lemma 3.2 and this proves Step 1.

Step 2 T *is a maximal torus in $\mathrm{G}^c_{x_\infty}$ and there exists an element $g_\infty \in \mathrm{G}^c$ such that $x_\infty = g_\infty^{-1}x_0$ and $\mathrm{T} = g_\infty^{-1}\mathrm{T}g_\infty$.*

By assumption, the complexified group orbit $\mathrm{G}^c(x_0)$ contains a critical point of the square of the moment map. Hence $x_\infty \in \mathrm{G}^c(x_0)$ by Theorem 13.1. Choose an element $g_0 \in \mathrm{G}^c$ such that

$$g_0x_0 = x_\infty.$$

Since T is a maximal torus in $\mathrm{G}^c_{x_0}$ it follows that $g_0\mathrm{T}g_0^{-1}$ is a maximal torus in

$$g_0\mathrm{G}^c_{x_0}g_0^{-1} = \mathrm{G}^c_{g_0x_0} = \mathrm{G}^c_{x_\infty}.$$

Moreover, since x_0 is μ-balanced and $\mathrm{T} \subset \mathrm{G}$, it follows from part (i) of Lemma 13.14 that $\mathrm{T} \subset \mathrm{G}_{x_0} \subset \mathrm{G}_{x_\infty} \subset \mathrm{G}^c_{x_\infty}$. Since T has the same dimension as $g_0\mathrm{T}g_0^{-1}$, this shows that T is another maximal torus in $\mathrm{G}^c_{x_\infty}$. Now it follows from the Cartan–Iwasawa–Malcev Theorem in [50, Thm 14.1.3] that any two maximal tori in any connected Lie group are conjugate. Apply this result to the identity component of $\mathrm{G}^c_{x_\infty}$ to obtain an element $g_1 \in \mathrm{G}^c_{x_\infty}$ such that

$$g_1g_0\mathrm{T}g_0^{-1}g_1^{-1} = \mathrm{T}.$$

Then we also have $g_1g_0x_0 = g_1x_\infty = x_\infty$ and so the element $g_\infty := (g_1g_0)^{-1} \in \mathrm{G}^c$ satisfies the requirements of Step 2.

Step 3 $\mu(x_\infty) \in \mathfrak{t}$.

Note that

$$L_{x_\infty}\mu(x_\infty) = 0, \qquad [\mu(x_\infty), \mathfrak{t}] = 0$$

by part (iii) of Lemma 13.14. Hence

$$\mathfrak{t}' := \mathfrak{t} + \mathbb{R}\mu(x_\infty) \subset \ker L_{x_\infty}$$

is an abelian Lie subalgebra and therefore the set

$$\mathrm{T}' := \overline{\exp(\mathfrak{t}')} \subset \mathrm{G}_{x_\infty}$$

is a torus. It contains T and so must be equal to T by Step 2. Hence

$$\mu(x_\infty) \in \mathfrak{t}' \subset \mathrm{Lie}(\mathrm{T}') = \mathfrak{t}$$

and this proves Step 3.

Step 4 *Let $g : \mathbb{R} \to \mathrm{G}^c$ be as in Step 1 and let $g_\infty \in \mathrm{G}^c$ be as in Step 2. Define*

$$\tau_0 := \Pi_{\mathrm{T}}(\mu(x_0))$$

and choose $u : \mathbb{R} \to \mathrm{G}$ and $\eta : \mathbb{R} \to \mathfrak{g}$ such that

$$g(t)\exp(\mathbf{i}\eta(t))u(t) = g_\infty \exp(\mathbf{i}t\tau_0) \qquad \text{for all } t \in \mathbb{R}. \tag{13.22}$$

Then the function $\mathbb{R} \to \mathbb{R} : t \mapsto |\eta(t)|$ is nonincreasing.

Recall that $\pi : \mathrm{G}^c \to M := \mathrm{G}^c/\mathrm{G}$ denotes the canonical projection and define the curves $\gamma : \mathbb{R} \to M$ and $\gamma_\infty : \mathbb{R} \to M$ by

$$\gamma(t) := \pi\big(g(t)\big), \qquad \gamma_\infty(t) := \pi\big(g_\infty \exp(\mathbf{i}t\tau_0)\big)$$

for $t \in \mathbb{R}$. We prove that these are both negative gradient flow lines of the Kempf–Ness function $\Phi_{x_0} : M \to \mathbb{R}$. For γ this follows directly from the definition and part (vi) of Theorem 4.3. For γ_∞ we use the fact that $\mu(x_\infty) \in \mathfrak{t}$ by Step 3, thus

$$\mu(x_\infty) = \Pi_{\mathrm{T}}(\mu(x_0)) = \tau_0$$

by part (ii) of Lemma 13.15, and so

$$\mu(x_\infty) = \tau_0 \in \mathfrak{t} \subset \ker L_{x_0} \subset \ker L_{x_\infty}$$

by part (ii) of Lemma 13.14. This implies

$$\begin{aligned}\big(g_\infty \exp(\mathbf{i}t\tau_0)\big)^{-1}\frac{d}{dt}g_\infty \exp(\mathbf{i}t\tau_0) &= \mathbf{i}\tau_0\\ &= \mathbf{i}\mu(x_\infty)\\ &= \mathbf{i}\mu(\exp(-\mathbf{i}t\tau_0)x_\infty)\\ &= \mathbf{i}\mu\big((g_\infty \exp(\mathbf{i}t\tau_0))^{-1}x_0\big).\end{aligned}$$

Thus γ and γ_∞ are negative gradient flow lines of the Kempf–Ness function Φ_{x_0} as claimed.

By Eq. (13.22) and part (ii) of Theorem C.1, the curve

$$[0,1] \to M : s \mapsto \pi\big(g(t)\exp(\mathbf{i}s\eta(t))\big)$$

is a geodesic joining $\gamma(t)$ to $\gamma_\infty(t)$. Since M has nonpositive sectional curvature by part (iv) of Theorem C.1, geodesics are unique, and hence

$$|\eta(t)| = d(\gamma(t), \gamma_\infty(t)) \qquad \text{for all } t \in \mathbb{R}.$$

Since the Kempf–Ness function is convex along geodesics by part (i) of Theorem 4.3, it follows from Lemma A.2 that the function $t \mapsto |\eta(t)|$ is nonincreasing. This proves Step 4.

Step 5 $x_\infty \in \mathrm{G}^c_\mathrm{T}(x_0)$.
Let $g : \mathbb{R} \to \mathrm{G}^c_\mathrm{T}$ be as in Step 1, let g_∞ be as in Step 2, and let

$$\tau_0 := \Pi_\mathrm{T}(\mu(x_0)) \in \mathfrak{t}$$

and $u : \mathbb{R} \to \mathrm{G}$ and $\eta : \mathbb{R} \to \mathfrak{g}$ be as in Step 4. Then

$$\tau_1 := g_\infty \tau_0 g_\infty^{-1} \in \mathfrak{t}. \tag{13.23}$$

By (13.22) and (13.23), we have

$$g(t)\exp(\mathbf{i}\eta(t))u(t) = g_\infty \exp(\mathbf{i}t\tau_0) = \exp(\mathbf{i}t\tau_1)g_\infty \tag{13.24}$$

for all $t \in \mathbb{R}$. Since the function $t \mapsto |\eta(t)|$ is nonincreasing by Step 4, there exists a sequence $t_i \to \infty$ such that the limits

$$\eta_\infty := \lim_{i\to\infty} \eta(t_i), \qquad u_\infty := \lim_{i\to\infty} u(t_i) \tag{13.25}$$

exist. Define

$$a_\infty := \exp(\mathbf{i}\eta_\infty)u_\infty. \tag{13.26}$$

Since $\tau_1 \in \mathfrak{t} \subset \ker L_{x_0}$, it follows from (13.24), (13.25), and (13.26) that

$$\begin{aligned}
x_\infty &= \lim_{i\to\infty} g(t_i)^{-1}x_0 \\
&= \lim_{i\to\infty} g(t_i)^{-1}\exp(\mathbf{i}t_i\tau_1)x_0 \\
&= \lim_{i\to\infty} \exp(\mathbf{i}\eta(t_i))u(t_i)g_\infty^{-1}x_0 \\
&= \exp(\mathbf{i}\eta_\infty)u_\infty x_\infty \\
&= a_\infty x_\infty.
\end{aligned}$$

Thus $a_\infty \in \mathrm{G}^c_{x_\infty}$ and hence

$$x_\infty = a_\infty x_\infty = a_\infty g_\infty^{-1} x_0.$$

Moreover,

$$\exp(\mathbf{i}\eta(t))u(t)g_\infty^{-1} = g(t)^{-1}\exp(\mathbf{i}t\tau_1) \in \mathrm{G}^c_\mathrm{T}$$

for all $t \in \mathbb{R}$ by (13.24) and Step 1. Hence, by (13.25) and (13.26), we have

$$a_\infty g_\infty^{-1} = \exp(\mathbf{i}\eta_\infty)u_\infty g_\infty^{-1} = \lim_{i\to\infty}\exp(\mathbf{i}\eta(t_i))u(t_i)g_\infty^{-1} \in \mathrm{G}^c_\mathrm{T}.$$

This proves Step 5 and Theorem 13.2.

□

Proof of the Generalized Székelyhidi Moment-Weight Inequality

The next goal is to establish the generalized Székelyhidi moment-weight inequality in Theorem 13.3. In the rational case it is due to Székelyhidi [72, Theorem 1.3.6]. We derive it as a corollary of the standard moment-weight inequality in Theorem 6.7 for the $\mathrm{G}_\mathrm{T}/\mathrm{T}$-action on X_T in Lemma 13.9.

Proof of Theorem 13.3 Let $x \in X$, let $\mathrm{T} \subset \mathrm{G}_x$ be a torus with the Lie algebra $\mathfrak{t}$, and let G_T and $\mathfrak{g}_\mathrm{T}$ be as in (13.10). Then, by (13.12), we have

$$\left|[\xi]_\mathrm{T}\right|^2_{\mathfrak{g}_\mathrm{T}/\mathfrak{t}} = |\xi|^2 - |\Pi_\mathrm{T}(\xi)|^2 = |\xi - \Pi_\mathrm{T}(\xi)|^2 \tag{13.27}$$

for all $\xi \in \mathfrak{g}_{\mathrm{T}}$. Moreover, if $\zeta = \xi + \imath\eta \in \mathscr{T}^c \cap \mathfrak{g}_{\mathrm{T}}^c \setminus \mathfrak{t}$, then by part (ii) of Lemma 13.9 we have $\Pi_{\mathrm{T}}(\mathrm{Im}(\zeta)) = 0$, which implies

$$\left|[\eta]_{\mathrm{T}}\right|_{\mathfrak{g}_{\mathrm{T}}/\mathfrak{t}} = |\eta|$$

and hence

$$\begin{aligned}
|[\zeta]_{\mathrm{T}^c}|_c^2 &= |[\xi]_{\mathrm{T}}|_{\mathfrak{g}_{\mathrm{T}}/\mathfrak{t}}^2 - |[\eta]_{\mathrm{T}}|_{\mathfrak{g}_{\mathrm{T}}/\mathfrak{t}}^2 \\
&= |\xi|^2 - |\eta|^2 - |\Pi_{\mathrm{T}}(\xi)|^2 \\
&= |\zeta|_c^2 - |\Pi_{\mathrm{T}}(\zeta)|^2 \\
&= |\zeta - \Pi_{\mathrm{T}}(\zeta)|_c^2 .
\end{aligned} \tag{13.28}$$

Combine (13.27) and (13.28) with Lemma 13.12 and Theorem 12.1 to obtain

$$\begin{aligned}
\sup_{\zeta \in \mathscr{T}^c \cap \mathfrak{g}_{\mathrm{T}}^c \setminus \mathfrak{t}} \frac{-w_{\mu,\mathrm{T}}(x,\zeta)}{\sqrt{|\zeta|_c^2 - |\Pi_{\mathrm{T}}(\zeta)|^2}} &= \sup_{\zeta \in \mathscr{T}^c \cap \mathfrak{g}_{\mathrm{T}}^c \setminus \mathfrak{t}} \frac{-w_{\mu_{\mathrm{T}}}(x,[\zeta]_{\mathrm{T}^c})}{|[\zeta]_{\mathrm{T}^c}|_c} \\
&= \sup_{\substack{\xi \in \mathfrak{g}_{\mathrm{T}} \setminus \mathfrak{t} \\ \exp(\xi) \in \mathrm{T}}} \frac{-w_{\mu_{\mathrm{T}}}(x,[\xi]_{\mathrm{T}})}{|[\xi]_{\mathrm{T}}|_{\mathfrak{g}_{\mathrm{T}}/\mathfrak{t}}} \\
&= \sup_{\substack{\xi \in \mathfrak{g}_{\mathrm{T}} \setminus \mathfrak{t} \\ \exp(\xi) \in \mathrm{T}}} \frac{-w_{\mu,\mathrm{T}}(x,\xi)}{|\xi - \Pi_{\mathrm{T}}(\xi)|} \\
&= \sup_{\substack{\xi \in \mathfrak{g}_{\mathrm{T}} \cap \mathfrak{t}^{\perp} \setminus \{0\} \\ \exp(\xi) \in \mathrm{T}}} \frac{-w_{\mu}(x,\xi)}{|\xi|}.
\end{aligned} \tag{13.29}$$

Here the first step follows from (13.28) and part (iv) of Lemma 13.12, the second from Theorem 12.1 for the $\mathrm{G}_{\mathrm{T}}/\mathrm{T}$-action on X_{T} with the moment map

$$\mu_{\mathrm{T}} : X_{\mathrm{T}} \to \mathfrak{g}_{\mathrm{T}}/\mathfrak{t}$$

in Lemma 13.9, the third from (13.27) and part (iv) of Lemma 13.12, and the last from part (ii) of Lemma 13.12 by replacing ξ with $\xi - \Pi_{\mathrm{T}}(\xi)$.

Since $\Pi_{\mathrm{T}}(\mu(hx)) = \Pi_{\mathrm{T}}(\mu(x))$ for all $h \in \mathrm{G}_{\mathrm{T}}^c$ by Lemma 13.13, we have

$$\begin{aligned}
|\mu_{\mathrm{T}}(hx)|_{\mathfrak{g}_{\mathrm{T}}/\mathfrak{t}} &= \sqrt{|\mu(hx)|^2 - |\Pi_{\mathrm{T}}(\mu(hx))|^2} \\
&= \sqrt{|\mu(hx)|^2 - |\Pi_{\mathrm{T}}(\mu(x))|^2} \\
&= |\mu(hx) - \Pi_{\mathrm{T}}(\mu(x))|
\end{aligned} \tag{13.30}$$

for all $h \in \mathrm{G}_\mathrm{T}^c$ by (13.13) and (13.27). Hence

$$\begin{aligned}\sup_{\zeta\in\mathscr{T}^c\cap\mathfrak{g}_\mathrm{T}^c\setminus\mathfrak{t}}\frac{-w_{\mu,\mathrm{T}}(x,\zeta)}{\sqrt{|\zeta|_c^2-|\Pi_\mathrm{T}(\zeta)|^2}} &= \sup_{\zeta\in\mathscr{T}^c\cap\mathfrak{g}_\mathrm{T}^c\setminus\mathfrak{t}}\frac{-w_{\mu_\mathrm{T}}(x,[\zeta]_{\mathrm{T}^c})}{|[\zeta]_{\mathrm{T}^c}|_c} \\ &\le \inf_{h\in\mathrm{G}_\mathrm{T}^c}|\mu_\mathrm{T}(hx)|_{\mathfrak{g}_\mathrm{T}/\mathfrak{t}} \\ &= \inf_{h\in\mathrm{G}_\mathrm{T}^c}\sqrt{|\mu(hx)|^2-|\Pi_\mathrm{T}(\mu(x))|^2} \\ &= \inf_{h\in\mathrm{G}^c}\sqrt{|\mu(hx)|^2-|\Pi_\mathrm{T}(\mu(x))|^2}.\end{aligned} \tag{13.31}$$

Here the first step follows from (13.29), the second step follows from Theorem 6.7, the third step follows from (13.30), and the last step follows from part (iv) of Lemma 13.15. That the supremum on the left is attained follows from Theorem 10.3, and that equality holds in (13.31) if and only if the left hand side is nonnegative follows from Corollary 12.7 for the $\mathrm{G_T/T}$-action on X_T with the moment map $\mu_\mathrm{T} : X_\mathrm{T} \to \mathfrak{g}_\mathrm{T}/\mathfrak{t}$ in Lemma 13.9. This proves the Székelyhidi moment-weight inequality (13.5) in the case $g = \mathbb{1}$.

To prove the result in general, replace G by $\widetilde{\mathrm{G}} := g^{-1}\mathrm{G}g$, with the inner product (13.1) on its Lie algebra $\widetilde{\mathfrak{g}} := g^{-1}\mathfrak{g}g$, and use the moment map $\widetilde{\mu} : X \to \widetilde{\mathfrak{g}}$ in (13.7) for the $\widetilde{\mathrm{G}}$-action on $(X, g^*\omega)$. Then Lemma 13.5 shows that the estimate (13.5) is equivalent to the same estimate with μ replaced by $\widetilde{\mu}$. This proves Theorem 13.3. □

In the rational case the inequality (13.5) in Theorem 13.3 is due to Székelyhidi. In [72, Thm 1.3.6] he stated the estimate in the following form.

Corollary 13.16 (Székelyhidi) *Let $x \in X$, let $\mathrm{T} \subset \mathrm{G}_x$ be a torus with the Lie algebra $\mathfrak{t}$, and let $\mathrm{G_T}$ and $\mathfrak{g}_\mathrm{T}$ be as in (13.10). Let $\xi \in \mathfrak{g} \setminus \mathfrak{t}$ such that $[\xi, \mathfrak{t}] = 0$ and $w_{\mu,\mathrm{T}}(x, \xi) < 0$. Then*

$$|\Pi_\mathrm{T}(\mu(x))|^2 + \frac{w_{\mu,\mathrm{T}}(x,\xi)^2}{|\xi|^2 - |\Pi_\mathrm{T}(\xi)|^2} \le \inf_{h\in\mathrm{G}^c}|\mu(hx)|^2. \tag{13.32}$$

Proof This follows by taking $\zeta = \xi$ and $g = \mathbb{1}$ in Theorem 13.3 and squaring the inequality (13.5). □

Proof of the Hilbert–Mumford Criterion for Critical Orbits

By Theorem 13.2 we can use the Hilbert–Mumford criterion for polystability in Theorem 12.5 to characterize the complexified group orbits that contain critical points of the square of the moment map.

Proof of Theorem 13.4 Assume first that $x \in X$ is μ-balanced, that $\mathrm{T} \subset \mathrm{G}_x$ is a maximal torus, and that $g = \mathbb{1}$. Then $x \in X_\mathrm{T}$. Consider the Hamiltonian $\mathrm{G_T/T}$-action on X_T in Lemma 13.9 with the moment $\mu_\mathrm{T} : X_\mathrm{T} \to \mathfrak{g}_\mathrm{T}/\mathfrak{t}$.

Claim 1 *Condition (i) in Theorem 13.4 holds if and only if x is μ_{T}-polystable.*

Claim 2 *Condition (ii) in Theorem 13.4 is equivalent to condition (ii) in Theorem 12.5 for the quadruple* $(X_{\mathrm{T}}, \mathrm{G}_{\mathrm{T}}/\mathrm{T}, \mu_{\mathrm{T}}, x)$.

Claim 3 *Condition (iii) in Theorem 13.4 is equivalent to condition (v) in Theorem 12.5 for the quadruple* $(X_{\mathrm{T}}, \mathrm{G}_{\mathrm{T}}/\mathrm{T}, \mu_{\mathrm{T}}, x)$.

Theorem 13.2 with $g = \mathbb{1}$ asserts that $\mathrm{G}^c(x)$ contains a critical point of the square of the moment map if and only if there exists an $h \in \mathrm{G}^c_{\mathrm{T}}$ such that

$$\mu(hx) \in \mathfrak{t}.$$

By Lemma 13.9 this holds if and only if x is μ_{T}-poystable. This proves Claim 1. Claim 3 follows directly from part (iv) of Lemma 13.12 and the characterization of toral generators in Lemma 13.11.

We prove Claim 2. Every equivalence class in $\mathfrak{g}_{\mathrm{T}}/\mathfrak{t}$ has a unique representative

$$\xi \in \mathfrak{g}_{\mathrm{T}} \cap \mathfrak{t}^{\perp}.$$

The equivalence class $[\xi]_{\mathrm{T}} \in \mathfrak{g}_{\mathrm{T}}/\mathfrak{t}$ is nonzero if and only if $\xi \neq 0$, and it satisfies $\exp([\xi]_{\mathrm{T}}) = [\mathbb{1}]_{\mathrm{T}} \in \mathrm{G}_{\mathrm{T}}/\mathrm{T}$ if and only if $\exp(\xi) \in \mathrm{T}$. Thus our lattice vector in $\mathfrak{g}_{\mathrm{T}}/\mathfrak{t}$ can be represented by a unique element $\xi \in \mathscr{T}^c$ that satisfies the conditions

$$[\xi, \mathfrak{t}] = 0, \qquad \exp(\xi) \in \mathrm{T}, \qquad \xi \in \mathfrak{g}, \qquad \Pi_{\mathrm{T}}(\xi) = 0.$$

Since

$$w_{\mu_{\mathrm{T}}}(x, [\xi]_{\mathrm{T}}) = w_{\mu,\mathrm{T}}(x, \xi) = w_{\mu}(x, \xi)$$

for any such ξ, by part (iv) of Lemma 13.12, this proves Claim 2.

Under the assumption $\mathrm{T} \subset \mathrm{G}_x$, the assertions of Theorem 13.4 follow directly from Claim 1, Claim 2, Claim 3, and Theorem 12.5.

Now let T be a maximal torus in G^c_x and let $g \in \mathrm{G}^c$ such that $g\mathrm{T}g^{-1} \subset \mathrm{G}$. Consider the subgroup $\widetilde{\mathrm{G}} := g^{-1}\mathrm{G}g \subset \mathrm{G}^c$ and the moment map $\widetilde{\mu} : X \to \widetilde{\mathfrak{g}}$ in (13.7). Then $\mathrm{G}^c(x)$ contains a critical point of $\frac{1}{2}|\mu|^2$ if and only if it contains a critical point of $\frac{1}{2}|\widetilde{\mu}|^2$. Moreover, $w_{\widetilde{\mu},\mathrm{T}}(x, \zeta) = w_{\mu,\mathrm{T}}(x, \zeta)$ for all $\zeta \in \mathscr{T}^c \cap \mathfrak{g}^c_{\mathrm{T}} \setminus \mathfrak{t}$ by Lemma 13.5. Thus each of the conditions (i), (ii), (iii) in Theorem 13.4 for $(X, \omega, \mathrm{G}, \mu, g, \mathrm{T}, x)$ with $\mathrm{T} \subset \mathrm{G}^c_x$ is equivalent to the corresponding condition for $(X, \widetilde{\omega}, \widetilde{\mathrm{G}}, \widetilde{\mu}, \mathbb{1}, \mathrm{T}, x)$ with $\mathrm{T} \subset \widetilde{\mathrm{G}}_x$, for which the equivalence has already been established. This proves Theorem 13.4. □

The Rational Case

If the inner product on $\mathfrak{g}$ is rational with some factor $\hbar > 0$ (Definition 9.9), then for every torus $T \subset \mathrm{G}$ with the Lie algebra $\mathfrak{t} = \mathrm{Lie}(\mathrm{T})$ there exists a unique connected Lie subgroup $\mathrm{H} \subset \mathrm{G}$ with the Lie algebra $\mathfrak{h} = \mathfrak{g}_{\mathrm{T}} \cap \mathfrak{t}^{\perp}$ (Lemma 13.17). In this situation the action of the quotient group $\mathrm{G_T}/\mathrm{T}$ on X_T can be replaced by the action of the subgroup H on all of X and the Székelyhidi criterion in Theorem 13.2 can be restated in terms of μ_H-polystability (Corollary 13.18).

Lemma 13.17 (Székelyhidi [72, Lemma 1.3.2]) *Assume the inner product on* $\mathfrak{g}$ *is rational with factor* $\hbar$. *Let* $\mathrm{T} \subset \mathrm{G}$ *be a torus with the Lie algebra* $\mathfrak{t} := \mathrm{Lie}(\mathrm{T})$, *let* $\mathrm{G_T}$ *and* $\mathfrak{g}_\mathrm{T}$ *be as in* (13.10), *and define*

$$\mathfrak{h} := \left\{ \xi \in \mathfrak{g} \,\middle|\, [\xi, \tau] = 0 \text{ and } \langle \xi, \tau \rangle = 0 \text{ for all } \tau \in \mathfrak{t} \right\},$$

$$\mathrm{H} := \left\{ u(1) \,\middle|\, \begin{array}{l} u : [0,1] \to \mathrm{G} \text{ is a smooth path} \\ \text{such that } u(0) = \mathbb{1} \text{ and} \\ \dot{u}(t)u(t)^{-1} \in \mathfrak{h} \text{ for all } t \in [0,1] \end{array} \right\}. \tag{13.33}$$

Then H *is a Lie subgroup of* G *with the Lie algebra* $\mathrm{Lie}(\mathrm{H}) = \mathfrak{h}$. *Moreover,*

$$\mathrm{G_T} := \mathrm{TH} = \left\{ av \,\middle|\, a \in \mathrm{T},\ v \in \mathrm{H} \right\}, \qquad \mathfrak{g}_\mathrm{T} = \mathfrak{t} \oplus \mathfrak{h}, \tag{13.34}$$

and the complexifications are related by

$$\mathrm{G}_\mathrm{T}^c = \mathrm{T}^c\mathrm{H}^c, \qquad \mathfrak{g}_\mathrm{T}^c = \mathfrak{t}^c \oplus \mathfrak{h}^c.$$

Proof The linear subspace $\mathfrak{h} \subset \mathfrak{g}$ in (13.33) is a Lie subalgebra and hence H is a subgroup of G (called the *integral subgroup of* $\mathfrak{h}$). We must prove that it is closed. By a theorem of Malcev [60] the subgroup H is closed if and only if $\overline{\{\exp(t\eta) \,|\, t \in \mathbb{R}\}} \subset \mathrm{H}$ for all $\eta \in \mathfrak{h}$ (see also Hilgert–Neeb [50, Cor 14.5.6]).

Fix an element $\eta \in \mathfrak{h}$. Then $\exp(t\eta)$ commutes with every element of $\mathfrak{t}$ and hence the linear subspace $\mathfrak{t} + \mathbb{R}\eta$ is an abelian subalgebra of $\mathfrak{g}$. Hence

$$T_\eta := \overline{\exp(\mathfrak{t} + \mathbb{R}\eta)} \subset \mathrm{G}$$

is a torus. Denote

$$\mathfrak{t}_\eta := \mathrm{Lie}(\mathrm{T}_\eta) \subset \mathfrak{g}.$$

Then the lattice $\mathfrak{t} \cap \Lambda$ spans $\mathfrak{t}$, and the lattice $\mathfrak{t}_\eta \cap \Lambda$ spans $\mathfrak{t}_\eta$. Consider the orthogonal decomposition

$$\mathfrak{t}_\eta = \mathfrak{t} \oplus \mathfrak{t}', \qquad \mathfrak{t}'_\eta := \mathfrak{t}_\eta \cap \mathfrak{t}_\eta^\perp.$$

We claim that the intersection $\mathfrak{t}'_\eta \cap \Lambda$ spans $\mathfrak{t}'_\eta$. To see this, define

$$m := \dim(\mathfrak{t}), \qquad n := \dim(\mathfrak{t}_\eta),$$

choose an integral basis $e_1, \dots, e_m$ of $\mathfrak{t} \cap \Lambda$, and extend it to an integral basis $e_1, \dots, e_n$ of $\mathfrak{t}_\eta \cap \Lambda$.

Use Gram–Schmidt to obtain an orthogonal basis $e'_1, \dots, e'_n$ of $\mathfrak{t}_\eta$ defined recursively by $e'_1 := e_1$ and

$$e'_k := e_k - \sum_{i=1}^{k-1} \frac{\langle e_k, e'_i \rangle}{|e'_i|^2} e'_i \qquad \text{for } k = 2, \dots, n.$$

It follows by induction that e'_k is a rational linear combination of $e_1, \dots, e_k$ for each k and thus satisfies

$$|e'_k|^2 \in 2\pi\hbar\mathbb{Q}, \qquad \langle e_j, e'_k \rangle \in 2\pi\hbar\mathbb{Q}$$

for all j. Since

$$\mathrm{span}(e'_1, \dots, e'_k) = \mathrm{span}(e_1, \dots, e_k)$$

for all k, the vectors $e'_{m+1}, \dots, e'_n$ form a rational basis of $\mathfrak{t}'_\eta \cap \mathbb{Q}\Lambda$, so $\mathfrak{t}'_\eta \cap \Lambda$ spans $\mathfrak{t}'_\eta$ as claimed. Thus

$$\mathrm{T}'_\eta := \exp(\mathfrak{t}'_\eta)$$

is a closed subgroup of G such that

$$\exp(\mathbb{R}\eta) \subset \mathrm{T}'_\eta \subset \mathrm{H}$$

and so $\overline{\exp(\mathbb{R}\eta)} \subset \mathrm{T}'_\eta \subset \mathrm{H}$. This shows that

$$\overline{\exp(\mathbb{R}\eta)} \subset \mathrm{H} \qquad \text{for all } \eta \in \mathfrak{h}.$$

Hence it follows from Malcev's theorem [60] that H is a closed subgroup of G and so is a Lie subgroup of G.

To prove (13.34), observe that

$$\mathfrak{g}_\mathrm{T} = \mathfrak{t} \oplus \mathfrak{h}, \qquad \mathrm{TH} \subset \mathrm{G}_\mathrm{T}$$

by definition. To prove the converse inclusion, choose a smooth path $u : [0, 1] \to \mathrm{G}$ such that

$$u(0) = \mathbb{1}, \qquad \xi(t) := u(t)^{-1}\dot{u}(t) \in \mathfrak{g}_\mathrm{T} \text{ for } 0 \leq t \leq 1.$$

Then $\xi(t) \in \mathfrak{t} \oplus \mathfrak{h}$ for all t and we write

$$\xi(t) = \tau(t) + \eta(t), \qquad \tau(t) \in \mathfrak{t}, \qquad \eta(t) \in \mathfrak{h}.$$

Define the curves $a : [0, 1] \to \mathrm{T}$ and $v : [0, 1] \to \mathrm{H}$ by

$$a^{-1}\dot{a} = \tau, \qquad v^{-1}\dot{v} = \eta, \qquad a(0) = v(0) = \mathbb{1}.$$

Then $u(t) = a(t)v(t)$ for all t, because T and H commute. Thus $\mathrm{G_T} \subset \mathrm{TH}$ and so

$$\mathrm{G_T} = \mathrm{TH}.$$

The same argument shows that $\mathrm{G}_T^c = \mathrm{T}^c\mathrm{H}^c$ and this proves Lemma 13.17. □

The following corollary is the Székelyhidi criterion in its original form, as stated in [72, Theorem 1.3.4].

Corollary 13.18 (Székelyhidi Criterion) *Assume that the inner product on $\mathfrak{g}$ is rational with factor $\hbar$ and that $x_0 \in X$ is μ-balanced. Let $\mathrm{T} \subset \mathrm{G}_{x_0}$ be a maximal torus with the Lie algebra $\mathfrak{t}$, let* H *and $\mathfrak{h}$ be as in Lemma 13.17, and let $\Pi_\mathrm{H} : \mathfrak{g} \to \mathfrak{h}$ be the orthogonal projection. Then $\mu_\mathrm{H} := \Pi_\mathrm{H} \circ \mu : X \to \mathfrak{h}$ is a moment map for the action of* H *on X and the following are equivalent.*

(i) $\mathrm{G}^c(x_0)$ *contains a critical point of the square of the moment map.*
(ii) x_0 *is μ_H-polystable.*
(iii) *Every $\zeta \in \mathscr{T}^c \cap \mathfrak{h}^c$ satisfies $w_\mu(x_0, \zeta) \geq 0$ and*

$$w_\mu(x_0, \zeta) = 0 \qquad \Longrightarrow \qquad \lim_{t\to\infty} \exp(\mathbf{i}t\zeta)x_0 \in \mathrm{H}^c(x_0). \tag{13.35}$$

(iv) *Every $\zeta \in \Lambda^c \cap \mathfrak{h}^c$ satisfies $w_\mu(x_0, \zeta) \geq 0$ and* (13.35).
(v) *Every $\zeta = \xi \in \mathfrak{h} \setminus \{0\}$ satisfies $w_\mu(x_0, \xi) \geq 0$ and* (13.35).
(vi) *Every $\zeta = \xi \in \Lambda \cap \mathfrak{h}$ satisfies $w_\mu(x_0, \xi) \geq 0$ and* (13.35).

Proof Assertion (ii) is equivalent to condition (ii) in Theorem 13.2 because

$$\mathrm{G}_\mathrm{T}^c = \mathrm{H}^c\mathrm{T}^c$$

by Lemma 13.17. Hence the equivalence of (i) and (ii) follows from Theorem 13.2. That (ii) is also equivalent to the remaining assertions was proved in Theorem 12.5. This proves Corollary 13.18. □

Chapter 14
Examples

Example 14.1 (Circle Actions) Let (X, ω) be a closed symplectic manifold and let $H : X \to \mathbb{R}$ be a smooth function whose Hamiltonian flow is 2π-periodic. Denote by $v_H \in \mathrm{Vect}(X)$ the Hamiltonian vector field of H and by $\{\varphi_H^s\}_{s\in\mathbb{R}}$ the flow of v_H. Thus

$$\iota(v_H)\omega = dH, \qquad \frac{d}{ds}\varphi_H^s = v_H \circ \varphi_H^s, \qquad \varphi_H^0 = \mathrm{id}, \qquad \varphi_H^{s+2\pi} = \varphi_H^s$$

for all $s \in \mathbb{R}$. Then the map $S^1 \times X \to X : (e^{\mathbf{i}s}, x) \mapsto \varphi_H^s(x)$ is a Hamiltonian circle action on X and $\mu := \mathbf{i}H : X \to \mathbf{i}\mathbb{R} = \mathrm{Lie}(S^1)$ is an equivariant moment map for this action. Choose an S^1-invariant and ω-compatible almost complex structure J on X. Then $\langle\cdot,\cdot\rangle := \omega(\cdot, J\cdot)$ is a Riemannian metric on X, the gradient of H with respect to this Riemannian metric is given by $\nabla H = Jv_H$, and the equation $\mathscr{L}_{v_H} J = 0$ implies

$$[v_H, \nabla H] = 0.$$

Hence the flows induced by the vector fields v_H and ∇H commute and so the circle action extends to a $\mathbb{C}^*$-action $\mathbb{C}^* \times X \to X : (g, x) \mapsto g \cdot x$ via

$$e^{\mathbf{i}(s+\mathbf{i}t)} \cdot x := \varphi_H^s \circ \varphi_{\nabla H}^t(x) \tag{14.1}$$

for $s, t \in \mathbb{R}$ and $x \in X$, where $\mathbb{R} \to \mathrm{Diff}(X) : t \mapsto \varphi_{\nabla H}^t$ denotes the (positive) gradient flow of H. Thus, for each $x_0 \in X$, the curve $x : \mathbb{R} \to X$, defined by

$$x(t) := e^{-t} \cdot x_0 = \varphi_{\nabla H}^t(x_0)$$

© The Author(s), under exclusive license to Springer Nature Switzerland AG 2021

V. Georgoulas et al., *The Moment-Weight Inequality and the Hilbert–Mumford Criterion*, Lecture Notes in Mathematics 2297,
https://doi.org/10.1007/978-3-030-89300-2_14

for $t \in \mathbb{R}$, is the positive gradient flow line of H through x_0, i.e.

$$\dot{x}(t) = \nabla H(x(t)), \qquad x(0) = x_0. \tag{14.2}$$

Denote the limit points of this gradient flow line by

$$x^+ := \lim_{t \to \infty} x(t), \qquad x^- := \lim_{t \to -\infty} x(t). \tag{14.3}$$

In spite of the fact that J need not be integrable, and so (X, ω, J) need not be a Kähler manifold, all the definitions and results in this book carry over to the present situation. For example, the weights of x_0 are

$$w_\mu(x_0, \mathbf{i}) = H(x^+), \qquad w_\mu(x_0, -\mathbf{i}) = -H(x^-). \tag{14.4}$$

Thus the weights are all nonnegative if and only if

$$H(x^-) \le 0 \le H(x^+),$$

and this holds if and only if the closure

$$\overline{\mathbb{C}^* \cdot x_0} = \mathbb{C}^* \cdot x_0 \cup \{x^-, x^+\}$$

of the complexified group orbit intersects the zero set $H^{-1}(0)$ of the moment map, i.e. x_0 is μ-semistable. If x_0 is μ-polystable, but not μ-stable, then the criterion of Theorem 12.5 translates into the condition $x^- = x_0 = x^+$, in which case the gradient flow line (14.2) is constant and so

$$H(x^-) = 0 = H(x^+).$$

The point x_0 is μ-stable if and only if its complexified group orbit $\mathbb{C}^* \cdot x_0$ intersects $H^{-1}(0)$ and the gradient flow line (14.2) is nonconstant. This holds if and only if

$$H(x^-) < 0 < H(x^+),$$

confirming the criterion of Theorem 12.6. In this situation the Kempf–Ness function $\Phi_{x_0} : (0, \infty) \cong \mathbb{C}^*/S^1 \to \mathbb{R}$ is given by

$$\Phi_{x_0}(r) = \int_1^r \frac{H(\varphi_{\nabla H}^{\log(\rho)}(x_0))}{\rho}\, d\rho \qquad \text{for } r > 0. \tag{14.5}$$

The metric on the space $M = \mathbb{C}^*/S^1 \cong (0, \infty)$ is given by $r^{-1}dr$ and the geodesics have the form $\gamma(t) = r_0 e^{ct}$. Convexity of Φ_{x_0} along geodesics translates into the equation

$$\frac{d^2}{dt^2}\Phi_{x_0}(e^t) = \left|\nabla H(\varphi^t_{\nabla H}(x_0))\right|^2 \geq 0.$$

Note, however, that

$$r^2\Phi''_{x_0}(r) = |\nabla H(\varphi^{\log(r)}_{\nabla H}(x_0))|^2 - H(\varphi^{\log(r)}_{\nabla H}(x_0))$$

and thus the function $\Phi_{x_0} : (0, \infty) \to \mathbb{R}$ need not be convex. The Generalized Székelyhidi Criterion for critical orbits in Theorem 13.2 asserts, in the present case, that the complexified group orbit $\mathbb{C}^* \cdot x_0$ contains a critical point of the function $\frac{1}{2}H^2 : X \to \mathbb{R}$ if and only if x_0 is either μ-stable (the case $\mathrm{T} = \{1\}$) or is a fixed point of the circle action (the case $\mathrm{T} = S^1$).

Example 14.2 (A Circle Action on S^2) Consider the unit sphere

$$S^2 := \left\{ x = (x_1, x_2, x_3) \in \mathbb{R}^3 \,\middle|\, |x|^2 = \sum_{i=1}^{3} x_i^2 = 1 \right\} \tag{14.6}$$

equipped with the standard symplectic and complex structures given by

$$\sigma_x(\widehat{x}, \widehat{y}) := \langle x \times \widehat{x}, \widehat{y} \rangle, \qquad J(x)\widehat{x} := x \times \widehat{x}$$

for $x \in S^2$ and $\widehat{x}, \widehat{y} \in T_x S^2 = x^\perp$. The standard circle action is given by rotation in the (x_1, x_2)-plane and is generated by the Hamiltonian function $H_c(x) := x_3 + c$ for $x \in S^2$. The complexified group action $\mathbb{C}^* \times S^2 \to S^2 : (g, x) \mapsto gx$ is given by

$$g \cdot x = \frac{1}{(1 + x_3) + |g|^2(1 - x_3)} \begin{pmatrix} 2(\mathrm{Re}(g)x_1 - \mathrm{Im}(g)x_2) \\ 2(\mathrm{Re}(g)x_2 + \mathrm{Im}(g)x_1) \\ (1 + x_3) - |g|^2(1 - x_3) \end{pmatrix} \tag{14.7}$$

for $g \in \mathbb{C}^*$ and $x \in S^2$. This corresponds to the standard $\mathbb{C}^*$-action on the Riemann sphere $\overline{\mathbb{C}} = \mathbb{C} \cup \{\infty\}$ under the stereographic projection

$$S^2 \to \overline{\mathbb{C}} : x \mapsto \frac{x_1 + \mathbf{i}x_2}{1 + x_3}, \tag{14.8}$$

which sends the north pole

$$N := (0, 0, 1)$$

to $z = 0$ and the south pole

$$S := (0, 0, -1)$$

to $z = \infty$. The Kempf–Ness function $\Phi_x : (0, \infty) \to \mathbb{R}$ of an element $x \in S^2$ is given by the explicit formula

$$\Phi_x(r) = (c-1)\log(r) + \log\left(\frac{|z|^2 + r^2}{|z|^2 + 1}\right), \qquad z := \frac{x_1 + \mathbf{i}x_2}{1 + x_3}. \tag{14.9}$$

The slopes are

$$w_\mu(x, \mathbf{i}) = \lim_{t\to\infty} \frac{\Phi(e^t)}{t} = \begin{cases} c+1, & \text{if } x \neq S, \\ c-1, & \text{if } x = S, \end{cases}$$

$$w_\mu(x, -\mathbf{i}) = \lim_{t\to\infty} \frac{\Phi(e^{-t})}{t} = \begin{cases} -(c-1), & \text{if } x \neq N, \\ -(c+1), & \text{if } x = N. \end{cases}$$

This is consistent with (14.4), Lemma 5.2, and the Hilbert–Mumford criterion. Namely, when $-1 < c < 1$ then the elements of $S^2 \setminus \{S, N\}$ are stable while S and N are unstable, and when $|c| > 1$ all elements of S^2 are unstable. If $c = 1$ then S is polystable, the elements of $S^2 \setminus \{S, N\}$ are all semistable, and N is unstable. If $c = -1$ then N is polystable, the elements of $S^2 \setminus \{S, N\}$ are all semistable, and S is unstable.

Example 14.3 (The SO(3)*-action on* S^2*)* The group $\mathrm{G} := \mathrm{SO}(3)$ acts on the unit sphere $S^2 \subset \mathbb{R}^3$ and the action preserves the standard Kähler structure in Example 14.2. Throughout we identify the Lie algebra $\mathfrak{so}(3)$ with $\mathbb{R}^3$ via the isomorphism

$$\mathbb{R}^3 \to \mathfrak{so}(3) : \xi = (\xi_1, \xi_2, \xi_3) \mapsto A_\xi := \begin{pmatrix} 0 & -\xi_3 & \xi_2 \\ \xi_3 & 0 & -\xi_1 \\ -\xi_2 & \xi_1 & 0 \end{pmatrix}. \tag{14.10}$$

This isomorphism identifies the cross product with the Lie bracket, satisfies the equation $A_{u\xi} = uA_\xi u^{-1}$ for $u \in \mathrm{SO}(3)$ and $\xi \in \mathbb{R}^3$, and the standard inner product on $\mathbb{R}^3$ is invariant and coincides with half the trace, i.e. $\langle \xi, \eta \rangle = -\frac{1}{2}\mathrm{trace}(A_\xi A_\eta)$ for all $\xi, \eta \in \mathbb{R}^3$. The SO(3)-action on S^2 is Hamiltonian and, under the above identification of the Lie algebra with $\mathbb{R}^3$, the moment map $\mu : S^2 \to \mathbb{R}^3$ is the canonical inclusion, i.e. $\mu(x) = x$.

Throughout we identify SO(3) with $\mathrm{SU}(2)/\{\pm\mathbb{1}\}$ via the isomorphism induced by the Lie group homomorphism $\mathrm{SU}(2) \to \mathrm{SO}(3)$ given by

$$\begin{pmatrix} a & b \\ -\bar{b} & \bar{a} \end{pmatrix} \mapsto \begin{pmatrix} \mathrm{Re}(a^2 - b^2) & -\mathrm{Im}(a^2 - b^2) & \mathrm{Re}(2ab) \\ \mathrm{Im}(a^2 - b^2) & \mathrm{Re}(a^2 + b^2) & \mathrm{Im}(2ab) \\ -\mathrm{Re}(2\bar{a}b) & -\mathrm{Im}(2\bar{a}b) & |a|^2 - |b|^2 \end{pmatrix}. \tag{14.11}$$

The homomorphism (14.11) has been chosen such that, under the stereographic projection $S^2 \to \overline{\mathbb{C}}$ in (14.8), the SO(3)-action on S^2 corresponds to the standard SU(2)-action on the Riemann sphere $\overline{\mathbb{C}}$ via

$$\begin{pmatrix} a & b \\ -\bar{b} & \bar{a} \end{pmatrix} \cdot z = \frac{az + b}{-\bar{b}z + \bar{a}} \tag{14.12}$$

(see [41, Proposition 4.2.2]). Thus

$$\mathrm{G}^c = \mathrm{PSL}(2, \mathbb{C})$$

is the group of Möbius transformations, it acts on the Riemann sphere by

$$\begin{pmatrix} a & b \\ c & d \end{pmatrix} \cdot z = \frac{az + b}{cz + d}, \tag{14.13}$$

and the corresponding action on S^2 via the stereographic projection (14.8) is given by

$$\begin{pmatrix} a & b \\ c & d \end{pmatrix} \cdot x = \frac{1}{|az + b|^2 + |cz + d|^2} \begin{pmatrix} 2\mathrm{Re}\big((az + b)\overline{(cz + d)}\big) \\ 2\mathrm{Im}\big((az + b)\overline{(cz + d)}\big) \\ |az + b|^2 - |cz + d|^2 \end{pmatrix} \tag{14.14}$$

for $x \in S^2$ and $z := (1 + x_3)^{-1}(x_1 + \mathbf{i}x_2)$.

To obtain an explicit formula for the Kempf–Ness function, it is convenient to identify the symmetric space G^c/G with the hyperbolic 3-space

$$\mathbb{H}^3 = \left\{ y = (y_0, y_1, y_2, y_3) \in \mathbb{R}^4 \mid y_0 > 0,\ Q(y, y) = -1 \right\}, \tag{14.15}$$

where $Q : \mathbb{R}^4 \times \mathbb{R}^4 \to \mathbb{R}$ is the quadratic form

$$Q(y, y') := -y_0 y_0' + y_1 y_1' + y_2 y_2' + y_3 y_3'$$

for $y, y' \in \mathbb{R}^4$ and the Riemannian metric on $\mathbb{H}^3$ is given by the restriction of Q to the tangent spaces. The isometry $\mathscr{F} : \mathrm{G}^c/\mathrm{G} \to \mathbb{H}^3$ sends (the equivalence class of) the matrix $g \in \mathrm{SL}(2,\mathbb{C})$ to the vector $(y_0, y_1, y_2, y_3) \in \mathbb{H}^3$ given by

$$\begin{aligned} y_0 &:= (a+d)^2/2 - 1 \\ y_1 &:= (a+d)\mathrm{Re}(b), \\ y_2 &:= (a+d)\mathrm{Im}(b), \\ y_3 &:= (d^2-a^2)/2, \end{aligned} \qquad \begin{pmatrix} a & b \\ \bar{b} & d \end{pmatrix} := (gg^*)^{1/2}. \tag{14.16}$$

With this identification the lifted Kempf–Ness function $\Phi_x : \mathrm{PSL}(2,\mathbb{C}) \to \mathbb{R}$ associated to an element $x \in S^2$ is given by

$$\Phi_x(g) = \log\left(y_0 - x_1 y_1 - x_2 y_2 - x_3 y_3\right) \tag{14.17}$$

for $g \in \mathrm{PSL}(2,\mathbb{C})$, where $y \in \mathbb{H}^3$ is given by (14.16) (see [41, Thms 4.2.5 and 4.2.6]). In this example the μ-weight of a pair $(x,\xi) \in S^2 \times \mathbb{R}^3$ is

$$w_\mu(x,\xi) = \begin{cases} |\xi|, & \text{if } \xi \neq -|\xi|x, \\ -|\xi|, & \text{if } \xi = -|\xi|x. \end{cases} \tag{14.18}$$

The vector $\xi \in \mathbb{R}^3 \setminus \{0\}$ corresponds to the skew-Hermitian matrix

$$\widehat{u}_\xi := \frac{1}{2}\begin{pmatrix} \mathbf{i}\xi_3 & \xi_2 - \mathbf{i}\xi_1 \\ -\xi_2 - \mathbf{i}\xi_1 & -\mathbf{i}\xi_3 \end{pmatrix} \in \mathfrak{su}(2) \tag{14.19}$$

via the deriviative of the homomorphism (14.11) and so the geodesic ray γ_ξ in $\mathbb{H}^3$ is given by $\gamma_\xi(t) = \mathscr{F}(\exp(\mathbf{i}t\widehat{u}_\xi)) = (\cosh(t|\xi|), \sinh(t|\xi|)\frac{\xi}{|\xi|})$ for $t \geq 0$ (see [41, Lemma 4.2.8]). The Kempf–Ness function along this ray is

$$\Phi_x(\gamma_\xi)(t) = \log\left(\cosh(t|\xi|) - \sinh(t|\xi|)\frac{\langle \xi, x\rangle}{|\xi|}\right)$$

and the asymptotic slope of this function is the weight $w_\mu(x,\xi)$ in (14.18). The explicit computation shows that the weight function $\xi \mapsto w_\mu(x,\xi)$ is discontinuous. Geometrically, the asymptotic slope of the Kempf–Ness function (14.17) is -1 along precisely one geodesic ray of unit speed, and is $+1$ along all others. This corresponds to the fact that all elements $x \in S^2$ are μ-unstable because $|\mu| \equiv 1$.

Example 14.4 (The $\mathrm{SO}(3)$*-action on* $(S^2)^n$*)* Fix a positive integer n and real numbers $\lambda_i > 0$ for $i = 1, \dots, n$. Let $\sigma \in \Omega^2(S^2)$ be the standard symplectic form in Example 14.2 and consider the symplectic form

$$\omega_\lambda := \lambda_1\sigma \oplus \lambda_2\sigma \oplus \cdots \oplus \lambda_n\sigma \in \Omega^2((S^2)^n)$$

on $X := (S^2)^n$. This is a Kähler form for the standard complex structure J. The diagonal action of $\mathrm{G} = \mathrm{SO}(3)$ on $(S^2)^n$ preserves the Kähler structure (ω_λ, J) and is generated by the moment map $\mu_\lambda : (S^2)^n \to \mathbb{R}^3 \cong \mathfrak{so}(3)$, which assigns to each n-tuple $x = (x_1, \dots, x_n) \in (S^2)^n$ its center of mass

$$\mu_\lambda(x_1, \dots, x_n) = \lambda_1 x_1 + \lambda_2 x_2 + \cdots + \lambda_n x_n.$$

(See Example 14.3.) The negative gradient flow of the moment map squared is given by the solutions of the differential equation $\dot{x} = -JL_x\mu_\lambda(x)$ which in the present situation takes the form $\dot{x}_i = -x_i \times (\mu_\lambda(x) \times x_i) = \langle \mu_\lambda(x), x_i \rangle x_i - \mu_\lambda(x)$, i.e.

$$\dot{x}_i = \left\langle \sum_{j=1}^n \lambda_j x_j, x_i \right\rangle x_i - \sum_{j=1}^n \lambda_j x_j, \qquad i = 1, \dots, n. \tag{14.20}$$

An element $x = (x_1, \dots, x_n) \in (S^2)^n$ is a critical point of the moment map squared (i.e. satisfies $L_x\mu_\lambda(x) = 0$) if and only if either $\mu_\lambda(x) = 0$ or the vectors $x_1, \dots, x_n$ are parallel. By the computations in Example 14.3 the Kempf–Ness function of an element $x = (x_1, \dots, x_n) \in (S^2)^n$ is given by

$$\Phi_x(g) = \sum_{i=1}^n \lambda_i \log(y_0 - x_{i1}y_1 - x_{i2}y_2 - x_{i3}y_3) \tag{14.21}$$

for $g \in \mathrm{PSL}(2, \mathbb{C})$, where $y \in \mathbb{H}^3$ is given by (14.16). The μ_λ-weight of a pair (x, ξ) with $x = (x_1, \dots, x_n) \in (S^2)^n$ and $\xi \in \mathbb{R}^3 \setminus \{0\}$ is given by

$$\frac{w_{\mu_\lambda}(x, \xi)}{|\xi|} = \sum_{i=1}^n \lambda_i \frac{w_\mu(x_i, \xi)}{|\xi|} = \sum_{x_i \neq -\xi/|\xi|} \lambda_i - \sum_{x_i = -\xi/|\xi|} \lambda_i. \tag{14.22}$$

Fix an element $x = (x_1, \dots, x_n) \in (S^2)^n$. Then the weighted sum of the Dirac measures at the points x_i defines a function $\nu_x : S^2 \to [0, \nu]$ given by

$$\nu := \sum_{i=1}^n \lambda_i, \qquad \nu_x(p) := \sum_{x_i = p} \lambda_i \qquad \text{for } p \in S^2. \tag{14.23}$$

Define

$$\kappa^+(x) := \max_{i=1,\dots,n} \nu_x(x_i), \qquad \kappa^-(x) := \min_{i=1,\dots,n} \nu_x(x_i). \tag{14.24}$$

With these formulas at hand the stability properties of an element $x \in (S^2)^n$ are determined by the Hilbert–Mumford criterion as follows.

When $\xi \in \mathbb{R}^3$ has norm $|\xi| = 1$, it follows from (14.22) and (14.23) that

$$w_{\mu_\lambda}(x, \xi) = \nu - 2\nu_x(-\xi). \tag{14.25}$$

By Theorem 12.3 an element $x \in (S^2)^n$ is μ_λ-unstable if and only if it admits a negative weight, and according to (14.25) this means that more than half the mass is concentrated at a single element of S^2. Thus, by (14.24),

$$x \text{ is } \mu_\lambda\text{-unstable} \qquad \Longleftrightarrow \qquad \kappa^+(x) > \frac{\nu}{2}. \tag{14.26}$$

Among the μ_λ-unstable points are those where the mass is concentrated at precisely two elements of S^2 with greater mass at one of these points. Those are the elements whose G^c-orbits contain higher critical points of the moment map squared. It follows directly from (14.26) that

$$x \text{ is } \mu_\lambda\text{-semistable} \qquad \Longleftrightarrow \qquad \kappa^+(x) \le \frac{\nu}{2}. \tag{14.27}$$

If $\kappa^+(x) = \nu/2$ then precisely half the mass is located at one element of S^2. The other half of the mass is also located at a single point if and only if the element is μ_λ-polystable but not μ_λ-stable, because any pair of points on the sphere is equivalent to an antipodal pair via a Möbius transformation. Thus

$$x \text{ is } \mu_\lambda\text{-polystable and not } \mu_\lambda\text{-stable} \qquad \Longleftrightarrow \qquad \kappa^+(x) = \kappa^-(x) = \frac{\nu}{2}. \tag{14.28}$$

Using the fact that for $\xi, x_i \in S^2$ we have $\lim_{t\to\infty} \exp(\mathbf{i}t\xi)x_i = \xi$ whenever $x_i \neq -\xi$ and $\lim_{t\to\infty} \exp(\mathbf{i}t\xi)x_i = -\xi$ whenever $x_i = -\xi$, one can verify that this is consistent with the Hilbert–Mumford criterion in Theorem 12.5. Finally, it follows from Theorem 12.6 and (14.25) that

$$x \text{ is } \mu_\lambda\text{-stable} \qquad \Longleftrightarrow \qquad \kappa^+(x) < \frac{\nu}{2}. \tag{14.29}$$

In this case at least three points on S^2 have positive mass and so the isotropy subgroup is trivial. The Hilbert–Mumford criterion asserts in this case that there exists a Möbius transformation $g \in \mathrm{PSL}(2, \mathbb{C})$ such that the weighted center of mass of the points $gx_1, \ldots, gx_n$ is zero.

This example does not satisfy the rationality conditions of Chap. 9 whenever the λ_i are rationally independent. The three notions of μ_λ-semistability, μ_λ-polystability, and μ_λ-stability are equivalent whenever n is odd and $\lambda_i = 1$. In this case the quotient space $\mathscr{M}_n := X^{\mathrm{s}}/\mathrm{G}^c = \mu^{-1}(0)/\mathrm{G} = X/\!\!/\mathrm{G}$ is a smooth manifold, called the **Mumford quotient**, and can be viewed as compactification of the configuration space $\mathscr{C}_n := ((S^2)^n \setminus \Delta_n)/\mathrm{PSL}(2, \mathbb{C})$, where $\Delta_n \subset (S^2)^n$ denotes the fat diagonal. A finer compactification is the Deligne–Mumford space $\overline{\mathscr{M}}_{0,n}$ and there is a natural projection $\pi : \overline{\mathscr{M}}_{0,n} \to \mathscr{M}_n$ which sends a stable curve of genus zero with n marked points to its *Mumford component* (see [61, page 650]).

Example 14.5 (Normal Matrices) This example is taken from a lecture by Peter Kronheimer [54]. Consider the vector space

$$V := \mathbb{C}^{n\times n}$$

with the Hermitian inner product

$$\langle A, B\rangle := \mathrm{Re}(\mathrm{trace}(A^*B))$$

for $A, B \in \mathbb{C}^{n\times n}$ and the projective space $X := \mathbb{P}(V)$ with the Fubini–Study form

$$\omega_A(\widehat{A}_1, \widehat{A}_2) := \frac{\langle \mathbf{i}\widehat{A}_1, \widehat{A}_2\rangle}{|A|^2} - \frac{\langle \mathbf{i}\widehat{A}_1, A\rangle\langle \widehat{A}_2, A\rangle}{|A|^4} + \frac{\langle \widehat{A}_1, A\rangle\langle \mathbf{i}\widehat{A}_2, A\rangle}{|A|^4} \tag{14.30}$$

for $A \in V \setminus \{0\}$ and $\widehat{A}_i \in V$. The special unitary group

$$\mathrm{G} := \mathrm{SU}(n)$$

acts on V by

$$u \cdot A := uAu^{-1}$$

for $u \in \mathrm{SU}(n)$ and $A \in V$. The induced action on the projective space $\mathbb{P}(V)$ is Hamiltonian and generated by the moment map $\mu : \mathbb{P}(V) \to \mathfrak{su}(n)$ whose lift to $V \setminus \{0\}$ is given by

$$\mu(A) = -\frac{\mathbf{i}}{2}\frac{[A, A^*]}{|A|^2} \tag{14.31}$$

for $A \in V \setminus \{0\}$. Here the Lie algebra $\mathfrak{g} := \mathfrak{su}(n)$ is equipped with the standard inner product

$$\langle \xi, \eta\rangle := \mathrm{Re}(\mathrm{trace}(\xi^*\eta))$$

for $\xi, \eta \in \mathfrak{su}(n)$ and the infinitesimal covariant action of $\mathfrak{g}$ on $\mathbb{P}(V)$ is given by

$$L_A\xi := [\xi, A] - \frac{\mathrm{trace}([\xi, A]A^*)}{|A|^2}A \tag{14.32}$$

for $\xi \in \mathfrak{su}(n)$ and $A \in V \setminus \{0\}$. The right hand side in (14.32) is the projection of the commutator $[\xi, A]$ onto the complex orthogonal complement of A. Equation (14.32) implies (14.31). The complexified group action is given by

$$g \cdot A = gAg^{-1}$$

for $g \in \mathrm{SL}(n, \mathbb{C})$ and $A \in \mathbb{C}^{n\times n}$. A matrix $A \in \mathbb{C}^{n\times n} \setminus \{0\}$ belongs to the zero set of the moment map if and only if it is normal. It is μ-unstable if and only if it has a trivial spectrum $\sigma(A) = \{0\}$, is μ-semistable if and only if $\sigma(A) \neq \{0\}$, and is μ-polystable if and only if it is diagonalizable (and hence has at least one nonzero eigenvalue). In the case $n \geq 2$ there are no μ-stable points because the isotropy subgroup is always nontrivial.

The stability conditions in this example are characterized by the Mumford μ-weights as in Lemma 8.4. Namely, if $\xi \in \mathfrak{su}(n)$ and

$$\lambda_1 < \lambda_2 < \cdots < \lambda_k$$

are the eigenvalues of the Hermitian matrix $\mathbf{i}\xi$ with the associated eigenspace decomposition $\mathbb{C}^n = E_1 \oplus E_2 \oplus \cdots \oplus E_k$, then A is a block matrix with $A_{ij} : E_j \to E_i$ and

$$\mathbf{i}[\xi, A] = \begin{pmatrix} 0 & (\lambda_1 - \lambda_2)A_{12} & \cdots & (\lambda_1 - \lambda_k)A_{1k} \\ (\lambda_2 - \lambda_1)A_{21} & 0 & & (\lambda_2 - \lambda_k)A_{2k} \\ \vdots & & \ddots & \vdots \\ (\lambda_k - \lambda_1)A_{k1} & (\lambda_k - \lambda_2)A_{k2} & \cdots & 0 \end{pmatrix}.$$

Thus it follows from part (i) of Lemma 8.4 (with $\hbar = \frac{1}{2}$) that

$$w_\mu(A, \xi) = \frac{1}{2} \max_{A_{ij} \neq 0} (\lambda_i - \lambda_j).$$

If this number is negative for some ξ then the block matrix $(A_{ij})_{i,j=1,\dots,k}$ is strictly upper triangular and so $\sigma(A) = \{0\}$. The converse follows by considering the Jordan normal form and using the invariance of the weights under the action of the complexified group (Theorem 5.3). The moment-weight (in)equality in this example takes the form

$$\sup_{\xi \in \mathfrak{su}(n)\setminus\{0\}} \frac{\min_{A_{ij}\neq 0}(\lambda_j - \lambda_i)}{\sqrt{\sum_i \lambda_i^2 \dim^c(E_i)}} = \inf_{g \in \mathrm{SL}(n,\mathbb{C})} \frac{|[g^{-1}Ag, g^*A^*g^{*-1}]|}{|g^{-1}Ag|^2}, \tag{14.33}$$

where the λ_i and E_i are determined by ξ as above. By Corollary 12.7 equality holds in (14.33) because there are no stable points.

The square of the moment map is the function $f : \mathbb{P}(V) \to \mathbb{R}$ given by

$$f(A) = \frac{1}{2}|\mu(A)|^2 = \frac{|[A, A^*]|^2}{8|A|^4} \tag{14.34}$$

for $A \in V \setminus \{0\}$ and its negative gradient flow lines are the solutions of the differential equation

$$\dot{A} = -\frac{[[A, A^*], A]}{2|A|^2} + \frac{|[A, A^*]|^2}{2|A|^4} A. \tag{14.35}$$

Each solution of (14.35) satisfies $\mathrm{trace}(A^* \dot{A}) = 0$. By Lemma 3.2 it can be written in the form $A(t) = g(t)^{-1} A g(t)$, where $A := A(0)$ and the curve $g : \mathbb{R} \to \mathrm{SL}(n, \mathbb{C})$ satisfies the differential equation

$$g^{-1}\dot{g} = \frac{[g^{-1}Ag, g^*A^*(g^*)^{-1}]}{2|g^{-1}Ag|^2}, \qquad g(0) = \mathbb{1}. \tag{14.36}$$

The curve $g : \mathbb{R} \to \mathrm{SL}(n, \mathbb{C})$ in (14.36) is a negative gradient flow line of the lifted Kempf–Ness function $\Phi_A : \mathrm{SL}(n, \mathbb{C}) \to \mathbb{R}$ given by

$$\Phi_A(g) = \tfrac{1}{2} \log\left(\frac{|g^{-1}Ag|}{|A|}\right) \tag{14.37}$$

for $g \in \mathrm{SL}(n, \mathbb{C})$ (see Lemma 8.3). The quotient $\mathrm{G}^c/\mathrm{G} = \mathrm{SL}(n, \mathbb{C})/\mathrm{SU}(n)$ is isometrically isomorphic to the space $\mathscr{P}_n$ of positive definite Hermitian matrices with determinant one, equipped with the Riemannian metric

$$\langle \widehat{P}_1, \widehat{P}_2 \rangle_P := \tfrac{1}{4}\mathrm{trace}\big(P^{-1}\widehat{P}_1 P^{-1}\widehat{P}_2\big) \tag{14.38}$$

for $P \in \mathscr{P}_n$ and $\widehat{P}_i \in T_P\mathscr{P}_n$ (so $\widehat{P}_i$ is a Hermitian matrix with $\mathrm{trace}(P^{-1}\widehat{P}_i) = 0$). The isometry is given by $\mathrm{SL}(n, \mathbb{C})/\mathrm{SU}(n) \to \mathscr{P}_n : g \mapsto P := gg^*$. Under this isometry the Kempf–Ness function (14.37) is given by

$$\Phi_A(P) = \tfrac{1}{4} \log\left(\frac{\mathrm{trace}(APA^*P^{-1})}{\mathrm{trace}(A^*A)}\right) \tag{14.39}$$

for $P \in \mathscr{P}_n$ and the differential equation (14.36) translates into

$$\dot{P} = \frac{APA^* - PA^*P^{-1}AP}{\mathrm{trace}(PA^*P^{-1}A)}, \qquad P(0) = \mathbb{1}. \tag{14.40}$$

If A is diagonalizable and has nonzero spectrum (the polystable case) then the solution of (14.40) converges to a matrix $P \in \mathscr{P}_n$ as t tends to infinity by Theorem 4.3. This limit defines a Hermitian inner product

$$\langle z, z' \rangle_P := \mathrm{Re}(z^* P^{-1} z') \tag{14.41}$$

on $\mathbb{C}^n$ with respect to which the matrix A is normal, i.e.

$$[A, PA^*P^{-1}] = 0.$$

It is also interesting to examine the higher critical points of the moment map squared. These are the solutions of the equation

$$[[A, A^*], A] = \lambda A, \qquad \lambda = \frac{|[A, A^*]|^2}{|A|^2}. \tag{14.42}$$

The Kirwan–Ness inequality in Corollary 6.2 asserts that every solution A of Eq. (14.42) satisfies the inequality

$$\frac{|[A, A^*]|^2}{|A|^4} \le \frac{|[gAg^{-1}, (gAg^{-1})^*]|^2}{|gAg^{-1}|^4} \tag{14.43}$$

for all $g \in \mathrm{SL}(n, \mathbb{C})$.

In this example every unstable G^c-orbit contains a critical point of the square of the moment map. Examples of solutions of (14.42) are matrices of the form

$$A := A_{d_1,\dots,d_m} := \begin{pmatrix} A_{d_1} & 0 & \cdots & 0 \\ 0 & \ddots & \ddots & \vdots \\ \vdots & \ddots & \ddots & 0 \\ 0 & \cdots & 0 & A_{d_m} \end{pmatrix} \tag{14.44}$$

with $\sum_j d_j = n$, $\max_j d_j \ge 2$, $A_1 = 0$, and

$$A_d := \begin{pmatrix} 0 & a_1 & 0 & \cdots & 0 \\ 0 & 0 & a_2 & \ddots & \vdots \\ \vdots & \ddots & \ddots & \ddots & 0 \\ \vdots & & \ddots & \ddots & a_{d-1} \\ 0 & \cdots & \cdots & 0 & 0 \end{pmatrix}, \qquad a_i := \sqrt{\frac{i(d-i)}{2}}, \tag{14.45}$$

for $d \ge 2$. This matrix satisfies

$$[A_d, A_d^*] = \begin{pmatrix} \frac{d-1}{2} & 0 & \cdots & \cdots & 0 \\ 0 & \frac{d-3}{2} & \ddots & & \vdots \\ \vdots & \ddots & \ddots & \ddots & \vdots \\ \vdots & & \ddots & \frac{3-d}{2} & 0 \\ 0 & \cdots & \cdots & 0 & \frac{1-d}{2} \end{pmatrix}.$$

Hence

$$[[A_d, A_d^*], A_d] = A_d$$

and

$$|A_d|^2 = \left|[A_d, A_d^*]\right|^2 = \frac{(d-1)d(d+1)}{12}.$$

Thus (14.42) holds with $\lambda = 1$ and

$$f(A_{d_1,\dots,d_m}) = \frac{3}{2\sum_{j=1}^m (d_j-1)d_j(d_j+1)}.$$

The matrix $A = A_{d_1,\dots,d_m}$ with

$$m = n-1, \qquad d_1 = 2, \qquad d_2 = \dots = d_{n-1} = 1,$$

satisfies

$$f(A) = \frac{1}{4}$$

and is an absolute maximum of f.

We examine the Szekelyhidi criterion in Corollary 13.8 for the matrix

$$A := \begin{pmatrix} J_{d_1} & 0 & \cdots & 0 \\ 0 & \ddots & \ddots & \vdots \\ \vdots & \ddots & \ddots & 0 \\ 0 & \cdots & 0 & J_{d_m} \end{pmatrix}, \qquad J_d := \begin{pmatrix} 0 & 1 & & 0 \\ \vdots & \ddots & \ddots & \\ \vdots & & \ddots & 1 \\ 0 & \cdots & \cdots & 0 \end{pmatrix} \in \mathbb{R}^{d\times d},$$

in Jordan normal form (with $\sum_j d_j = n$ and $\max_j d_j \geq 2$). This matrix is μ-balanced (Definition 13.6). Its isotropy subgroup contains a maximal torus T of dimension m consisting of diagonal matrices. The Lie algebra $\mathfrak{t} = \mathrm{Lie}(\mathrm{T})$ is the space of all diagonal matrices with trace zero whose diagonal entries $\xi_{11}, \dots, \xi_{1d_1}, \dots, \xi_{m1}, \dots, \xi_{md_m}$ are purely imaginary and satisfy the condition $\mathbf{i}\xi_{ji} - \mathbf{i}\xi_{j,i+1} = \lambda$ for all $j = 1, \dots, m$ and all $i = 1, \dots, d_j - 1$ and some real number λ. Every matrix commuting with $\mathfrak{t}$ is necessarily diagonal. Thus $\mathfrak{g}_{\mathrm{T}} \cap \mathfrak{t}^{\perp}$ generates a torus of dimension $n - m - 1$. Let $\eta \in \mathfrak{g}_{\mathrm{T}} \cap \mathfrak{t}^{\perp} \setminus \{0\}$ with the diagonal entries $\eta_{11}, \dots, \eta_{1d_1}, \dots, \eta_{m1}, \dots, \eta_{md_m}$. Then

$$\sum_{i=1}^{d_j} \eta_{ji} = 0 \text{ for } j = 1, \dots, m, \qquad \sum_{j=1}^{m}\sum_{i=1}^{d_j} i\eta_{ji} = 0.$$

Moreover, the weight $w_\mu(A, \eta)$ is the maximal entry of the matrix $[\mathbf{i}\eta, A]$ by Lemma 8.4. The nonzero entries are $\mathbf{i}\eta_{ji} - \mathbf{i}\eta_{j,i+1}$ with $1 \le j \le m$ and $1 \le i < d_j$, and at least one them is positive. Thus A is conjugate to a solution of (14.42) by Corollary 13.8. This also follows directly from the above discussion.

One can think of a matrix $A \in \mathbb{C}^{n\times n}$ as a translation invariant Cauchy–Riemann operator

$$\bar{\partial}_A = \bar{\partial} + A$$

on $\mathbb{C}$, associated to the Hermitian connection

$$\nabla = d + \Phi dx + \Psi dy, \qquad \Phi := \tfrac{1}{2}(A - A^*), \qquad \Psi := \tfrac{1}{2\mathbf{i}}(A + A^*),$$

so $A = \Phi + \mathbf{i}\Psi$. Then $\mathrm{SU}(n)$ is the group of translation invariant unitary gauge transformations and it acts on ∇ by conjugation $u^*\nabla = u^{-1} \circ \nabla \circ u$. The complexified action of $\mathrm{SL}(n, \mathbb{C})$ is given by $g^*\nabla = d + \widetilde{\Phi}\, dx + \widetilde{\Psi}\, dy$ with

$$\widetilde{\Phi} := \tfrac{1}{2}\Big(g^{-1}\Phi g + g^*\Phi(g^*)^{-1} + \mathbf{i}g^{-1}\Psi g - \mathbf{i}g^*\Psi(g^*)^{-1}\Big),$$
$$\widetilde{\Psi} := \tfrac{1}{2}\Big(g^{-1}\Psi g + g^*\Psi(g^*)^{-1} - \mathbf{i}g^{-1}\Phi g + \mathbf{i}g^*\Phi(g^*)^{-1}\Big)$$

for $g \in \mathrm{SL}(n, \mathbb{C})$ and $\Phi, \Psi \in \mathfrak{su}(n)$. The curvature of ∇ is the 2-form

$$F^\nabla = [\Phi, \Psi]dx \wedge dy = -\tfrac{\mathbf{i}}{2}[A, A^*]dx \wedge dy$$

and hence agrees with the moment map. Thus A is diagonalizable if and only if ∇ is gauge equivalent to a flat connection by a translation invariant complex gauge transformation.

Example 14.6 (Control Systems) Consider the action of the general linear group $\mathrm{G}^c := \mathrm{GL}(n, \mathbb{C})$ on the complex vector space $V := \mathbb{C}^{n\times n} \times \mathbb{C}^{n\times m} \times \mathbb{C}^{p\times n}$ by

$$g \cdot (A, B, C) := \big(gAg^{-1}, gB, Cg^{-1}\big)$$

for $g \in \mathrm{G}^c$ and $(A, B, C) \in V$. The action of the compact subgroup

$$\mathrm{G} := \mathrm{U}(n)$$

preserves the standard Hermitian inner product on V and the induced action on $\mathbb{P}(V)$ is Hamiltonian and generated by the standard moment map

$$\mu(A, B, C) = -\frac{\mathbf{i}}{2}\frac{[A, A^*] + BB^* - C^*C}{\mathrm{trace}(A^*A + B^*B + CC^*)} \tag{14.46}$$

for $(A, B, C) \in V \setminus \{0\}$. The control system (A, B, C) is μ-stable if and only if it is controllable and observable i.e.

$$\sum_{i=0}^{n-1} \operatorname{im} A^i B = \mathbb{C}^n, \qquad \bigcap_{i=0}^{n-1} \ker C A^i = \{0\} \tag{14.47}$$

(see [49]). Assume $\mu(A, B, C) = 0$ and the isotropy subgroup is discrete, i.e.

$$AA^* + BB^* = A^*A + C^*C, \tag{14.48}$$

$$[\xi, A] = 0, \quad \xi B = 0, \quad C\xi = 0 \qquad \Longrightarrow \qquad \xi = 0 \tag{14.49}$$

for all $\xi \in \mathfrak{g} := \mathfrak{u}(n)$. Let $z \in \mathbb{C}^n$ and $\lambda \in \mathbb{C}$ such that $Az = \lambda z$ and $Cz = 0$. Then

$$(A - \lambda)(A^* - \bar{\lambda})z + BB^*z = (A^* - \bar{\lambda})(A - \lambda)z + C^*Cz = 0$$

by (14.48) and hence $A^*z = \bar{\lambda}z$ and $B^*z = 0$. Thus the matrix $\xi := \mathbf{i}zz^* \in \mathfrak{g}$ commutes with A and satisfies $\xi B = 0$ and $C\xi = 0$, and so $\xi = 0$ by (14.49). This implies $\ker(A - \lambda) \cap \ker C = \{0\}$ and $\operatorname{im}(A - \lambda) + \operatorname{im} B = \mathbb{C}^n$ for all $\lambda \in \mathbb{C}$ and hence (A, B, C) satisfies (14.47).

Conversely, suppose that (A, B, C) is controllable and observable but not μ-stable. Then, by the Hilbert–Mumford criterion in Theorem 12.6, there exists a $\xi \in \mathfrak{g} \setminus \{0\}$ such that $w_\mu((A, B, C), \xi) \leq 0$. Let $\lambda_1 < \cdots < \lambda_k$ be the eigenvalues of $\mathbf{i}\xi$, let $\mathbb{C}^n = E_1 \oplus \cdots \oplus E_k$ be the corresponding eigenspace decomposition, and let $A_{ij} : E_j \to E_i$, $B_i : \mathbb{C}^m \to E_i$, $C_j : E_j \to \mathbb{C}^p$ be the corresponding components of the matrices A, B, C. Then it follows from Lemma 8.4 that $A_{ij} = 0$ for $i > j$ (so A is upper triangular), that $B_i = 0$ whenever $\lambda_i > 0$, and that $C_j = 0$ whenever $\lambda_j < 0$. Moreover, by controllability and observability we have $C_1 \neq 0$ and $B_k \neq 0$. In the case $k \geq 2$ we obtain $0 \leq \lambda_1 < \lambda_k \leq 0$, and in the case $k = 1$ we obtain $\lambda_1 = 0$ and so $\xi = 0$, a contradiction in both cases. This shows that the control system (A, B, C) is μ-stable if and only if it is controllable and observable.

In control theory the solutions of (14.48) are known as *balanced control systems* [49]. The relation between balanced control systems and geometric invariant theory was noted by Helmke and Moore. They also showed that the gradient flow of the Kempf–Ness function in this setting converges to a balanced control system. More precisely, the lifted Kempf–Ness function $\Phi_{A,B,C} : \mathrm{G}^c \to \mathbb{R}$ is given by

$$\Phi_{A,B,C}(g) = \tfrac{1}{4} \log \left(\frac{|g^{-1}Ag|^2 + |g^{-1}B|^2 + |Cg|^2}{|A|^2 + |B|^2 + |C|^2} \right) \tag{14.50}$$

and its gradient flow takes the form

$$g^{-1}\dot{g} = \frac{1}{2} \frac{[g^{-1}Ag, g^*A^*(g^*)^{-1}] + g^{-1}BB^*(g^*)^{-1} - g^*C^*Cg}{|g^{-1}Ag|^2 + |g^{-1}B|^2 + |Cg|^2}. \tag{14.51}$$

Under the isometry $G^c/G \to \mathscr{P}_n : [g] \mapsto P := gg^*$ in Example 14.5 the Kempf–Ness function is given by

$$\Phi_{A,B,C}(P) = \tfrac{1}{4}\log\left(\frac{\operatorname{trace}(APA^*P^{-1} + BB^*P^{-1} + PC^*C)}{\operatorname{trace}(A^*A + BB^* + C^*C)}\right) \tag{14.52}$$

for $P \in \mathscr{P}_n$ and its negative gradient flow equation has the form

$$\dot{P} = \frac{APA^* - PA^*P^{-1}AP + BB^* - PC^*CP}{\operatorname{trace}(APA^*P^{-1} + BB^*P^{-1} + PC^*C)}. \tag{14.53}$$

If (A, B, C) is controllable and observable, this flow converges to a positive definite Hermitian matrix $P \in \mathscr{P}_n$ by Theorem 4.3, and (A, B, C) is balanced with respect to the Hermitian structure (14.41) determined by P, i.e.

$$APA^*P^{-1} + BB^*P^{-1} = PA^*P^{-1}A + PC^*C. \tag{14.54}$$

The isotropy subgroup of a controllable and observable system (A, B, C) is trivial and thus the moduli space of conjugacy classes of such systems is a (noncompact) Kähler manifold. The moduli space of controllable pairs (A, B) fits into the GIT framework with $G^c = SL(n, \mathbb{C})$ and the balancing equation (14.48) replaced by

$$[A, A^*] + BB^* - \frac{\operatorname{trace}(BB^*)}{n}\mathbb{1} = 0. \tag{14.55}$$

The topology of these and related moduli spaces was studied by Uwe Helmke and his collaborators in the 1980s (see e.g. [45–49]).

In [3] Atiyah–Drinfeld–Hitchin–Manin used the solutions of the equations

$$[A_1, A_1^*] + [A_2, A_2^*] + BB^* - C^*C = 0, \qquad [A_1, A_2] + BC = 0, \tag{14.56}$$

with $B, C^* \in \mathbb{C}^{n\times m}$ and $A_1, A_2 \in \mathbb{C}^{n\times n}$ to construct anti-self-dual $SU(m)$-instantons of charge n on the four-sphere. This is the *ADHM construcion.*

Example 14.7 (Partial Flag Manifolds) Let $n_1, \dots, n_r$ be a finite sequence of positive integers and define $n := n_1 + \cdots + n_r$. A **partial flag of type** $(n_1, \dots, n_r)$ is a sequence of subspaces $\{0\} = F_0 \subset F_1 \subset \cdots \subset F_r = \mathbb{C}^n$ such that $\dim^c(F_i/F_{i-1}) = n_i$ for $i = 1, \dots, r$. The space $\mathscr{F}(n_1, \dots, n_r)$ of all partial flags of type $(n_1, \dots, n_r)$ can be identified with the quotient space

$$\mathscr{F}(n_1, \dots, n_r) \cong U(n)/(U(n_1) \times \cdots \times U(n_r)).$$

If $\lambda_1 > \lambda_2 > \cdots > \lambda_{r-1} > \lambda_r = 0$, then the centralizer of the matrix

$$\xi := \begin{pmatrix} \mathbf{i}\lambda_1 \mathbb{1}_{n_1} & 0 & \cdots & 0 \\ 0 & \mathbf{i}\lambda_2 \mathbb{1}_{n_2} & \ddots & \vdots \\ \vdots & \ddots & \ddots & 0 \\ 0 & \cdots & 0 & \mathbf{i}\lambda_r \mathbb{1}_{n_r} \end{pmatrix} \in \mathfrak{u}(n) \tag{14.57}$$

is the subgroup $C(\xi) = \mathrm{U}(n_1) \times \cdots \times \mathrm{U}(n_r)$ and so the map

$$\mathrm{U}(n)/(\mathrm{U}(n_1) \times \cdots \times \mathrm{U}(n_r)) \to \mathscr{O}(\xi) : [u] \mapsto u\xi u^*$$

gives rise to a diffeomorphism from the partial flag manifold $\mathscr{F}(n_1, \ldots, n_r)$ to the (co)adjoint orbit $\mathscr{O}(\xi) = \{u\xi u^* \mid u \in \mathrm{U}(n)\} \subset \mathfrak{u}(n)$. This orbit has the tangent space $T_\eta \mathscr{O}(\xi) = \mathrm{im}\,(\mathrm{ad}(\eta))$ at $\eta \in \mathscr{O}(\xi)$ and is equipped with a symplectic form $\omega_\eta(\widehat{\eta}_1, \widehat{\eta}_2) = \frac{1}{2}\mathrm{Re}\,\mathrm{trace}(\eta[\zeta_1, \zeta_2])$ for $\eta \in \mathscr{O}(\xi)$ and $\widehat{\eta}_i = [\eta, \zeta_i]$, $\zeta_i \in \mathfrak{u}(n)$.

The partial flag manifold $\mathscr{F}(n_1, \ldots, n_r)$ can be described as a symplectic quotient as follows. For $i = 1, \ldots, r$ define $k_i := n_1 + \cdots + n_i$ and consider the action of the group $\mathrm{G}^c := \mathrm{GL}(k_1, \mathbb{C}) \times \cdots \times \mathrm{GL}(k_{r-1}, \mathbb{C})$ on the complex vector space $V := \mathbb{C}^{k_2 \times k_1} \times \cdots \times \mathbb{C}^{k_r \times k_{r-1}}$ by

$$g \cdot A := \Big(g_2 A_1 g_1^{-1}, g_3 A_2 g_2^{-1}, \ldots, g_{r-1} A_{r-2} g_{r-2}^{-1}, A_{r-1} g_{r-1}^{-1}\Big)$$

for $g = (g_1, \ldots, g_{r-1}) \in \mathrm{G}^c$ and $A = (A_1, \ldots, A_{r-1}) \in V$. The space V is equipped with the symplectic form

$$\omega(A, B) = \sum_{i=1}^{r-1} \mathrm{Im}\,\mathrm{trace}(A_i^* B_i)$$

for $A = (A_1, \ldots, A_{r-1})$, $B = (B_1, \ldots, B_{r-1}) \in V$ with $A_i, B_i \in \mathbb{C}^{k_{i+1} \times k_i}$, and the action of the compact subgroup $G := \mathrm{U}(k_1) \times \cdots \times \mathrm{U}(k_{r-1})$ preserves this symplectic form. If we identify the Lie algebra $\mathfrak{g} := \mathrm{Lie}(\mathrm{G}) = \mathfrak{u}(k_1) \times \cdots \times \mathfrak{u}(k_{r-1})$ with its dual space $\mathfrak{g}^*$ via the inner product $\langle \xi, \eta \rangle = \sum_{i=1}^{r-1} \mathrm{Re}\,\mathrm{trace}(\xi_i^* \eta_i)$, then the action of G on V is generated by the moment map $\mu : V \to \mathfrak{g}$, given by

$$\mu(A) := \frac{\mathbf{i}}{2}\Big(A_1^* A_1, A_2^* A_2 - A_1 A_1^*, \ldots, A_{r-1}^* A_{r-1} - A_{r-2} A_{r-2}^*\Big)$$

for $A = (A_1, \ldots, A_{r-1}) \in V$.

Now fix a central element $\tau := \frac{\mathbf{i}}{2}(\tau_1 \mathbb{1}_{k_1}, \ldots, \tau_{r-1} \mathbb{1}_{k_{r-1}}) \in \mathfrak{g}$ such that $\tau_i > 0$ for $i = 1, \ldots, r-1$. Then, with $A_0 := 0$, we have

$$\mu^{-1}(\tau) = \big\{A \in V \,\big|\, A_i^* A_i - A_{i-1} A_{i-1}^* = \tau_i \mathbb{1}_{k_i} \text{ for } i = 1, \ldots, r-1\big\}.$$

The group $G = U(k_1) \times \cdots \times U(k_{r-1})$ acts freely on $\mu^{-1}(\tau)$ and the quotient

$$V /\!\!/ G(\tau) = \mu^{-1}(\tau)/G$$

is diffeomorphic to the partial flag manifold $\mathscr{F}(n_1, \ldots, n_r)$. The diffeomorphism sends the equivalence class of an element $A = (A_1, \ldots, A_{r-1}) \in \mu^{-1}(\tau)$ to the partial flag $F = (F_i)_{i=1,\ldots,r}$ given by $F_r = \mathbb{C}^n$ and

$$F_i := \operatorname{im}\left(A_{r-1} \circ A_{r-2} \circ \cdots \circ A_i : \mathbb{C}^{k_i} \to \mathbb{C}^n\right), \qquad i = 1, \ldots, r-1.$$

Moreover, if $\xi \in \mathfrak{u}(n)$ is chosen as in (14.57) with $\lambda_r = 0$ and

$$\lambda_i := \tau_i + \tau_{i+1} + \cdots + \tau_{r-1}, \qquad i = 1, \ldots, r-1,$$

then the map

$$V /\!\!/ G(\tau) = \mu^{-1}(\tau)/G \to \mathscr{O}(\xi) : [A] \mapsto \mathbf{i}A_{r-1}A_{r-1}^* \tag{14.58}$$

is a diffeomorphism. Indeed, if $A \in \mu^{-1}(\tau)$, then $\mathbf{i}\lambda_i$ is an eigenvalue o $\mathbf{i}A_{r-1}A_{r-1}^*$ with the n_i-dimensional eigenspace $E_i := \left\{A_{r-1}A_{r-2} \cdots A_i z \,\middle|\, z \in \mathbb{C}^{k_i},\ A_{i-1}^* z = 0\right\}$. Thus the image of the map (14.58) is contained in $\mathscr{O}(\xi)$. (**Exercise:** Prove that the map (14.58) is bijective and that it is a symplectomorphism.)

The lifted Kempf–Ness function $\Phi_A : G^c \to \mathbb{R}$ is given by

$$\Phi_A(g) = \tfrac{1}{4}\sum_{i=1}^{r-1}\left(\left|g_{i+1}^{-1}A_i g_i\right|^2 - |A_i|^2\right) - \tfrac{1}{4}\sum_{i=1}^{r-1}\tau_i \log\left(\det(g_i g_i^*)\right) \tag{14.59}$$

for $A \in V$ and $g \in G^c$. Here we define $g_r := \mathbb{1}_n$. The quotient space G^c/G can be identified with the space $\mathscr{P}$ of $(r-1)$-tuples $P = (P_1, \ldots, P_{r-1})$ of positive definite Hermitian matrices $P_i = P_i^* \in \mathbb{C}^{k_i \times k_i}$ via $[g_i] \mapsto P_i := g_i g_i^*$. In this formulation the Kempf-Ness function $\Phi_A : \mathscr{P} \to \mathbb{R}$ is given by

$$\Phi_A(P) = \tfrac{1}{4}\sum_{i=1}^{r-1}\left(\operatorname{trace}\left(A_i P_i A_i^* P_{i+1}^{-1}\right) - \operatorname{trace}(A_i^* A_i) - \tau_i \log\left(\det(P_i)\right)\right)$$

and its negative gradien flow wiith respect to the metric (14.38) has the form

$$\dot{P}_i = A_{i-1}P_{i-1}A_{i-1}^* - P_i A_i^* P_{i+1}^{-1} A_i P_i + \tau_i P_i \tag{14.60}$$

for $i = 1, \ldots, r-1$. Here we use the convention $P_r := \mathbb{1}_n$, $P_0 := 0$, $A_0 := 0$. An element $A \in V$ is $(\mu - \tau)$-stable if and only A_i has rank k_i for each i.

Example 14.8 (Lie Algebras) This example is due to Lauret [55]. Consider the standard action of the group $G^c = SL(n, \mathbb{C})$ on the complex vector space

$$V := \Lambda^2(\mathbb{C}^n)^* \otimes \mathbb{C}^n$$

of all complex bilinear 2-forms $\tau : \mathbb{C}^n \times \mathbb{C}^n \to \mathbb{C}^n$. The group action is given by

$$(g \cdot \tau)(z, z') := g\tau(g^{-1}z, g^{-1}z')$$

for $g \in SL(n, \mathbb{C})$, $\tau \in V$, $z, z' \in \mathbb{C}^n$. The action of the compact subgroup $G = SU(n)$ preserves the standard inner product on V, given by

$$\langle \sigma, \tau \rangle = \sum_{i,j=1}^{n} \langle \sigma(e_i, e_j), \tau(e_i, e_j) \rangle$$

for $\sigma, \tau \in V$. Here $\langle z, z' \rangle = \mathrm{Re} \sum_{i=1}^{n} \overline{z}_i z'_i$ denotes the standard Hermitian inner product on $\mathbb{C}^n$ and $e_1, \ldots, e_n$ is the standard basis of $\mathbb{C}^n$. By Lemma 8.2 the action of $SU(n)$ on the projective space $\mathbb{P}(V)$ with the symplectic form induced by the Hermitian structure on the sphere of radius $r = \sqrt{2\hbar}$ is generated by the moment map $\mu : \mathbb{P}(V) \to \mathfrak{su}(n)^*$ given by

$$\begin{aligned} \langle \mu(\tau), \xi \rangle &= \frac{\hbar}{|\tau|^2} \langle \tau, \mathbf{i}\xi \cdot \tau \rangle \\ &= \frac{\hbar}{|\tau|^2} \sum_{i,j=1}^{n} \langle \tau(e_i, e_j), \mathbf{i}\xi\tau(e_i, e_j) - 2\tau(e_i, \mathbf{i}\xi e_j) \rangle \end{aligned} \tag{14.61}$$

for $\tau \in V$ and $\xi \in \mathfrak{su}(n)$. Now suppose that $\tau : \mathbb{C}^n \times \mathbb{C}^n \to \mathbb{C}^n$ is a complex Lie bracket and denote its adjoint representation by $\mathrm{ad}_\tau : \mathbb{C}^n \to \mathrm{Der}(\mathbb{C}^n, \tau)$, so that $\mathrm{ad}_\tau(z) = \tau(z, \cdot)$ for $z \in \mathbb{C}^n$. Then, for all $\xi \in \mathfrak{su}(n)$,

$$\begin{aligned} \langle \mu(\tau), \xi \rangle &= -\frac{\mathbf{i}\hbar}{|\tau|^2} \mathrm{trace}(M_\tau \xi), \\ M_\tau &:= \sum_{i=1}^{n} \Big(2\mathrm{ad}_\tau(e_i)^* \mathrm{ad}_\tau(e_i) - \mathrm{ad}_\tau(e_i) \mathrm{ad}_\tau(e_i)^* \Big). \end{aligned} \tag{14.62}$$

Thus $\mu(\tau) = 0$ if and only if $M_\tau \in \mathbb{R}\mathbb{1}$. It was shown by Lauret in [55, Theorem 4.3] that the complexified group orbit $G^c([\tau]) \subset \mathbb{P}(V)$ intersects the zero set of the moment map (i.e. τ is polystable) if and only if the Lie algebra $(\mathbb{C}^n, \tau)$ is semisimple. His proof uses Cartan's theorem about the compact real form of a semisimple Lie algebra [10–12]. This theorem implies that in the semisimple case there exists a Hermitian inner product $\langle \cdot, \cdot \rangle'$ on $\mathbb{C}^n$ and a Hermitian orthonormal basis $e'_1, \ldots, e'_n$ of $\mathbb{C}^n$ such that $\mathrm{ad}_\tau(e'_i)^* = -\mathrm{ad}_\tau(e'_i)$ and $\sum_{i=1}^{n} \mathrm{ad}_\tau(e'_i)^2 = -c\mathbb{1}$ for some $c > 0$ and so $M'_\tau = c\mathbb{1}$. After suitable rescaling the two inner products are related by the action of $\mathrm{SL}(n, \mathbb{C})$ and so every semisimple complex Lie bracket τ is polystable.

One can modify this approach following the work of Donaldson in [36] so that Cartan's theorem does not need to be used but instead a proof of Cartan's theorem as well as various other standard results in Lie algebra theory emerge as byproducts of the proof. The key idea is to show directly that every simple Lie algebra is polystable by finding a critical point of the lifted Kempf–Ness function $\Phi_\tau : \mathrm{SL}(n, \mathbb{C}) \to \mathbb{R}$, given by

$$\Phi_\tau(g) := \hbar\Big(\log\left|g^{-1} \cdot \tau\right| - \log|\tau|\Big) = \frac{\hbar}{2}\log\left(\frac{|g^{-1} \cdot \tau|^2}{|\tau|^2}\right) \tag{14.63}$$

for $g \in \mathrm{SL}(n, \mathbb{C})$ (see Lemma 8.3). The Kempf–Ness function can be replaced by the function f_τ on the Hadamard manifold $\mathscr{P}_n \cong \mathrm{SL}(n, \mathbb{C})/\mathrm{SU}(n)$ of all positive definite Hermitian matrices of determinant one, defined by

$$f_\tau(gg^*) := \left|g^{-1} \cdot \tau\right|^2 \tag{14.64}$$

for $g \in \mathrm{SL}(n, \mathbb{C})$. The Hadamard manifold $\mathscr{P}_n$ can be identified with the space $\mathscr{H}_n$ of Hermitian inner products on $\mathbb{C}^n$ with a fixed determinant. Thus the task is to find a Hermitian inner product on $\mathbb{C}^n$ which minimizes the norm of the Lie bracket τ. This approach was suggested by Cartan [11] and carried out by Richardson [69] under the assumption that the Killing form is nondegenerate. In [36] Donaldson proved that every convex function $f : \mathscr{H}_n \to \mathbb{R}$ that is invariant under a subgroup of $\mathrm{SL}(n, \mathbb{C})$ that acts irreducibly on $\mathbb{C}^n$ has a critical point which is fixed by the subgroup. In the case at hand the relevant subgroup is the group $\mathrm{Aut}_0(\mathbb{C}^n, \tau) \cap \mathrm{SL}(n, \mathbb{C})$ of all Lie algebra automorphisms of $(\mathbb{C}^n, \tau)$ in the identity component with determinant one. It acts irreducibly on $\mathbb{C}^n$ whenever $(\mathbb{C}^n, \tau)$ is a simple Lie algebra. Every critical point of f_τ is then a Hermitian inner product h on $\mathbb{C}^n$ for which the space of derivations is invariant under the involution $A \mapsto A^*$. For Lie algebras with a trivial center the existence of such an inner product implies that the Killing form of $(\mathbb{C}^n, \tau)$ is nondegenerate, the adjoint representation $\mathrm{ad}_\tau : \mathbb{C}^n \to \mathrm{Der}(\mathbb{C}^n, \tau)$ is bijective, and $\mathrm{Aut}_0(\mathbb{C}^n, \tau)$ is actually contained in $\mathrm{SL}(n, \mathbb{C})$. Another byproduct of Donaldson's proof is that

$$\mathrm{K} := \mathrm{Aut}_0(\mathbb{C}^n, \tau) \cap \mathrm{SU}(\mathbb{C}^n, h)$$

is a maximal compact subgroup of $\mathrm{Aut}_0(\mathbb{C}^n, \tau)$, that every compact subgroup of $\mathrm{Aut}_0(\mathbb{C}^n, \tau)$ is conjugate to a subgroup of K, and that $\mathrm{Aut}_0(\mathbb{C}^n, \tau)$ is the complexification of K. Moreover, K is connected and the Lie algebra of K is Cartan's compact real form of $(\mathbb{C}^n, \tau)$. Once these results have been established for simple Lie algebras, it is a straight forward matter to deduce that the polystable points in the present setting are precisely the semisimple Lie algebras. For more details see [36, 55] and also [70, §7.6].

Appendix A
Nonpositive Sectional Curvature

Assume throughout that M is a nonempty, complete, connected, simply connected Riemannian manifold of nonpositive sectional curvature. Then Hadamard's theorem asserts that the exponential map

$$\exp_p : T_pM \to M$$

is a diffeomorphism for every $p \in M$ (see e.g. [70, Theorem 6.5.2]). The Levi-Civita connection on M is denoted by ∇ and the distance function associated to the Riemannian metric by $d : M \times M \to [0, \infty)$.

Lemma A.1 *Let $I \subset \mathbb{R}$ be an interval, let*

$$\gamma_0, \gamma_1 : I \to M$$

be smooth curves, and for each $t \in I$ let

$$[0, 1] \to M : s \mapsto \gamma(s, t)$$

be the unique geodesic with the endpoints

$$\gamma(0, t) = \gamma_0(t), \qquad \gamma(1, t) = \gamma_1(t).$$

Define the function $\rho : I \to \mathbb{R}$ by

$$\rho(t) := d(\gamma_0(t), \gamma_1(t)) = \int_0^1 |\partial_s \gamma(s, t)| \, ds. \tag{A.1}$$

© The Author(s), under exclusive license to Springer Nature Switzerland AG 2021

V. Georgoulas et al., *The Moment-Weight Inequality and the Hilbert–Mumford Criterion*, Lecture Notes in Mathematics 2297,
https://doi.org/10.1007/978-3-030-89300-2

If $t \in I$ such that $\rho(t) \neq 0$ then ρ is differentiable at t and

$$\begin{aligned}\dot\rho(t) &= \int_0^1 \frac{\langle \partial_s\gamma, \nabla_t\partial_s\gamma\rangle}{|\partial_s\gamma|}\,ds \\ &= \frac{\langle \partial_s\gamma(1,t), \partial_t\gamma(1,t)\rangle - \langle \partial_s\gamma(0,t), \partial_t\gamma(0,t)\rangle}{\rho(t)}.\end{aligned}$$

Proof If $\rho(t) \neq 0$ then $\partial_s\gamma(s,t) \neq 0$ for all s, so ρ is differentiable at t and

$$\begin{aligned}\dot\rho(t) &= \int_0^1 \partial_t|\partial_s\gamma|\,ds \\ &= \int_0^1 \frac{\langle \partial_s\gamma, \nabla_t\partial_s\gamma\rangle}{|\partial_s\gamma|}\,ds \\ &= \frac{1}{\rho(t)}\int_0^1 \partial_s\langle \partial_s\gamma, \partial_t\gamma\rangle\,ds.\end{aligned}$$

The last equation follows from the fact that $\nabla_s\partial_s\gamma \equiv 0$ and $|\partial_s\gamma(s,t)| = \rho(t)$. Now Lemma A.1 follows from the fundamental theorem of calculus. □

Lemma A.2 *Let $\Phi : M \to \mathbb{R}$ be a smooth function that is convex along geodesics, let $\gamma_0, \gamma_1 : \mathbb{R} \to M$ be negative gradient flow lines of Φ, and let γ and ρ be as in Lemma A.1. Then ρ is nonincreasing and, if*

$$\rho(t) \neq 0,$$

then

$$\dot\rho(t) = -\frac{1}{\rho(t)}\int_0^1 \frac{\partial^2}{\partial s^2}(\Phi\circ\gamma)(s,t)\,ds. \tag{A.2}$$

Proof Assume $\rho(t) \neq 0$. Then, by Lemma A.1, ρ is differentiable at t and

$$\begin{aligned}\dot\rho(t) &= \frac{1}{\rho(t)}\Big(\langle \partial_s\gamma(1,t), -\nabla\Phi(\gamma(1,t))\rangle - \langle \partial_s\gamma(0,t), -\nabla\Phi(\gamma(0,t))\rangle\Big) \\ &= -\frac{1}{\rho(t)}\Big(\partial_s(\Phi\circ\gamma)(1,t) - \partial_s(\Phi\circ\gamma)(0,t)\Big).\end{aligned}$$

This proves (A.2). If $\rho(t) = 0$ for some t then $\rho(t) = 0$ for all t. If $\rho(t) \neq 0$ for all t then ρ is nonincreasing by (A.2). This proves Lemma A.2. □

Lemma A.3 *Let $I \subset \mathbb{R}$ be an interval and let*

$$\gamma_0, \gamma_1 : I \to M$$

be smooth curves such that γ_0 is a geodesic. Define the function

$$\rho : I \to [0, \infty)$$

by (A.1).

(i) *If $\gamma_0(t) = \gamma_1(t)$ then*

$$\frac{d}{dt^{\pm}}\rho(t) = \lim_{\substack{h\to 0\\ h>0}} \frac{\rho(t \pm h)}{\pm h} = \pm|\dot\gamma_0(t) - \dot\gamma_1(t)|.$$

(ii) *If $\gamma_0(t) \neq \gamma_1(t)$ then*

$$\ddot\rho(t) \geq -|\nabla\dot\gamma_1(t)|.$$

Proof A theorem in differential geometry (e.g. [70, Lemma 4.7.1]) asserts that for every positive constant $\alpha < 1$ there exists a $\delta > 0$ such that, for all $v_0, v_1 \in T_pM$,

$$|v_0|, |v_1| < \delta \quad \Longrightarrow \quad \alpha|v_0 - v_1| \leq d(\exp_p(v_0), \exp_p(v_1)) \leq \alpha^{-1}|v_0 - v_1|.$$

Now assume $\rho(t_0) = 0$, denote $p_0 := \gamma_0(t_0) = \gamma_1(t_0)$, and choose a pair of smooth curves $v_0, v_1 : \mathbb{R} \to T_{p_0}M$ such that

$$\gamma_0(t_0 + t) = \exp_{p_0}(v_0(t)), \qquad \gamma_1(t_0 + t) = \exp_{p_0}(v_1(t))$$

for all t. Then

$$\begin{aligned}
\frac{d}{dt^+}\rho(t_0) &= \lim_{\substack{h\to 0\\ h>0}} \frac{\rho(t_0 + h)}{h} \\
&= \lim_{\substack{h\to 0\\ h>0}} \frac{d(\gamma_0(t_0 + h), \gamma_1(t_0 + h))}{h} \\
&= \lim_{\substack{h\to 0\\ h>0}} \frac{d(\exp_{p_0}(v_0(h)), \exp_{p_0}(v_1(h)))}{h} \\
&= \lim_{\substack{h\to 0\\ h>0}} \frac{|v_0(h) - v_1(h)|}{h} \\
&= \frac{d}{dt^+}\Big|_{t=0} |v_0(t) - v_1(t)|
\end{aligned}$$

$$
\begin{aligned}
&= |\dot{v}_0(0) - \dot{v}_1(0)| \\
&= |\dot{\gamma}_0(t_0) - \dot{\gamma}_1(t_0)|.
\end{aligned}
$$

An analogous argument shows that $\frac{d}{dt^-}\rho(t_0) = -|\dot{\gamma}_0(t_0) - \dot{\gamma}_1(t_0)|$. This proves the first assertion of Lemma A.3.

To prove the second assertion, define

$$
\gamma : [0, 1] \times I \to M, \qquad \rho : I \to \mathbb{R}
$$

as in Lemma A.1. If $\rho(t) \neq 0$ then Lemma A.1 asserts that ρ is differentiable at t and

$$
\begin{aligned}
\dot{\rho}(t) &= \int_0^1 \frac{\langle \partial_s \gamma, \nabla_t \partial_s \gamma \rangle}{|\partial_s \gamma|} \, ds \\
&= \frac{1}{\rho(t)} \Big(\langle \partial_s \gamma(1, t), \partial_t \gamma(1, t) \rangle - \langle \partial_s \gamma(0, t), \partial_t \gamma(0, t) \rangle \Big).
\end{aligned}
$$

In particular, by the Cauchy–Schwarz inequality, we have

$$
\dot{\rho}(t)^2 \leq \int_0^1 |\nabla_t \partial_s \gamma|^2 \, ds. \tag{A.3}
$$

Moreover,

$$
\frac{d}{dt}(\rho \dot{\rho}) = \rho \ddot{\rho} + \dot{\rho}^2
$$

and hence

$$
\rho(t)\ddot{\rho}(t) + \dot{\rho}(t)^2 = \frac{d}{dt}\Big(\langle \partial_s \gamma(1, t), \partial_t \gamma(1, t) \rangle - \langle \partial_s \gamma(0, t), \partial_t \gamma(0, t) \rangle \Big) = I + II,
$$

where

$$
I = \langle \partial_s \gamma(1, t), \nabla_t \partial_t \gamma(1, t) \rangle \geq -\rho(t) |\nabla_t \dot{\gamma}_1(t)|
$$

and

$$
\begin{aligned}
II &= \langle \nabla_t \partial_s \gamma(1, t), \partial_t \gamma(1, t) \rangle - \langle \nabla_t \partial_s \gamma(0, t), \partial_t \gamma(0, t) \rangle \\
&= \frac{\partial}{\partial s} \frac{|\partial_t \gamma(1, t)|^2}{2} - \frac{\partial}{\partial s} \frac{|\partial_t \gamma(0, t)|^2}{2} \\
&= \int_0^1 \frac{\partial^2}{\partial s^2} \frac{|\partial_t \gamma|^2}{2} \, ds
\end{aligned}
$$

$$
\begin{aligned}
&= \int_0^1 \Big(|\nabla_s \partial_t \gamma|^2 + \langle \partial_t \gamma, \nabla_s \nabla_s \partial_t \gamma \rangle \Big)\, ds \\
&= \int_0^1 \Big(|\nabla_t \partial_s \gamma|^2 + \langle \partial_t \gamma, \nabla_s \nabla_t \partial_s \gamma - \nabla_t \nabla_s \partial_s \gamma \rangle \Big)\, ds \\
&= \int_0^1 \Big(|\nabla_t \partial_s \gamma|^2 + \langle \partial_t \gamma, R(\partial_s \gamma, \partial_t \gamma) \partial_s \gamma \rangle \Big)\, ds \\
&\geq \dot\rho(t)^2.
\end{aligned}
$$

Here the last inequality follows from (A.3) and the fact that M has nonpositive sectional curvature. This proves Lemma A.3. □

Lemma A.4 *For all $p \in M$, all $v_0, v_1 \in T_pM$, and all $t \geq 1$,*

$$
|v_0 - v_1| \leq d(\exp_p(v_0), \exp_p(v_1)) \leq \frac{d(\exp_p(tv_0), \exp_p(tv_1))}{t}.
$$

Proof Define the curves $\gamma_0, \gamma_1 : [0, \infty) \to M$ and $\rho : [0, \infty) \to [0, \infty)$ by

$$
\gamma_0(t) := \exp_p(tv_0), \qquad \gamma_1(t) := \exp_p(tv_1), \qquad \rho(t) := d_M(\gamma_0(t), \gamma_1(t)).
$$

By Lemma A.3 the function ρ is convex and $\rho(0) = 0$ and $\dot\rho(0) = |v_0 - v_1|$. Hence

$$
d_M(\exp_p(tv_0), \exp_p(tv_1)) = \rho(t) \geq t\rho(1) = t d_M(\exp_p(v_0), \exp_p(v_1))
$$

for $t \geq 1$ and

$$
d_M(\exp_p(v_0), \exp_p(v_1)) = \rho(1) \geq \dot\rho(0) = |v_0 - v_1|.
$$

This proves Lemma A.4. □

Theorem A.5 (Cartan Fixed Point Theorem) *Let M be a nonempty, complete, connected, simply connected Riemannian manifold with nonpositive sectional curvature. Let* G *be a compact topological group that acts on M by isometries. Then there exists an element $p \in M$ such that*

$$
gp = p
$$

for every $g \in$ G.

Proof See page 157. □

The proof follows the argument given by Bill Casselmann in [13] and requires the following two lemmas. The first lemma asserts that every manifold of nonpositive sectional curvature is a semi-hyperbolic space in the sense of Alexandrov.

Lemma A.6 *Let M be a complete, connected, simply connected Riemannian manifold with nonpositive sectional curvature. Let $m \in M$ and $v \in T_mM$ and define*

$$p_0 := \exp_m(-v), \qquad p_1 := \exp_m(v).$$

Then

$$2d(m,q)^2 + \frac{d(p_0,p_1)^2}{2} \le d(p_0,q)^2 + d(p_1,q)^2$$

for every $q \in M$.

Proof By Hadamard's theorem the exponential map $\exp_m : T_mM \to M$ is a diffeomorphism (see e.g. [70, Theorem 6.5.2]). Hence

$$d(p_0,p_1) = 2|v|.$$

Now let $q \in M$. Then there exists a unique tangent vector $w \in T_mM$ such that

$$q = \exp_m(w), \qquad d(m,q) = |w|.$$

Since the exponential map is expanding, by Lemma A.4, we have

$$d(p_0,q) \ge |w+v|, \qquad d(p_1,q) \ge |w-v|.$$

Hence

$$\begin{aligned} d(m,q)^2 &= |w|^2 \\ &= \frac{|w+v|^2 + |w-v|^2}{2} - |v|^2 \\ &\le \frac{d(p_0,q)^2 + d(p_1,q)^2}{2} - \frac{d(p_0,p_1)^2}{4}. \end{aligned}$$

This proves Lemma A.6. □

The next lemma is **Serre's Uniqueness Theorem** for the *circumcentre* of a bounded set in a *semi-hyperbolic space*.

Lemma A.7 (Serre) *Let M be a nonempty, complete, connected, simply connected Riemannian manifold with nonpositive sectional curvature. For $p \in M$ and $r \ge 0$ denote by $B(p,r) \subset M$ the closed ball of radius r centered at p. Let $\Omega \subset M$ be a nonempty bounded set and define*

$$r_\Omega := \inf\{r > 0 \,|\, \textit{there exists a } p \in M \textit{ such that } \Omega \subset B(p,r)\}.$$

Then there exists a unique point $p_\Omega \in M$ such that $\Omega \subset B(p_\Omega, r_\Omega)$.

Proof We prove existence. Choose sequences $r_i > r_\Omega$ and $p_i \in M$ such that

$$\Omega \subset B(p_i, r_i), \qquad \lim_{i\to\infty} r_i = r_\Omega.$$

Choose $q \in \Omega$. Then $d(q, p_i) \le r_i$ for every i. Since the sequence r_i is bounded and M is complete, it follows that p_i has a convergent subsequence, still denoted by p_i. Its limit $p_\Omega := \lim_{i\to\infty} p_i$ satisfies $\Omega \subset B(p_\Omega, r_\Omega)$.

We prove uniqueness. Let $p_0, p_1 \in M$ such that

$$\Omega \subset B(p_0, r_\Omega) \cap B(p_1, r_\Omega).$$

Since the exponential map $\exp_p : T_pM \to M$ is a diffeomorphism there exists a unique tangent vector $v_0 \in T_{p_0}M$ such that $p_1 = \exp_{p_0}(v_0)$. Denote the midpoint between p_0 and p_1 by

$$m := \exp_{p_0}\left(\tfrac{1}{2}v_0\right).$$

Then it follows from Lemma A.6 that

$$d(m,q)^2 \le \frac{d(p_0,q)^2 + d(p_1,q)^2}{2} - \frac{d(p_0,p_1)^2}{4} \le r_\Omega^2 - \frac{d(p_0,p_1)^2}{4}$$

for every $q \in \Omega$. Since $\sup_{q\in\Omega} d(m,q) \ge r_\Omega$, by definition of the number r_Ω, it follows that $d(p_0, p_1) = 0$ and hence $p_0 = p_1$. This proves Lemma A.7. □

Proof of Theorem A.5 Fix an element $q \in M$ and consider the group orbit

$$\Omega := \{gq \mid g \in \mathrm{G}\}.$$

Since G is compact, this set is bounded. Let $r_\Omega \ge 0$ and $p_\Omega \in M$ be as in Lemma A.7. Then

$$\Omega \subset B(p_\Omega, r_\Omega).$$

Since G acts on M by isometries, this implies

$$\Omega = g\Omega \subset B(gp_\Omega, r_\Omega)$$

for all $g \in \mathrm{G}$. Hence it follows from the uniqueness statement in Lemma A.7 that $gp_\Omega = p_\Omega$ for every $g \in \mathrm{G}$. This proves Theorem A.5. □

Appendix B
The Complexified Group

Definition B.1 A **complex Lie group** is a Lie group G equipped with the structure of a complex manifold such that the structure maps

$$\mathrm{G} \times \mathrm{G} \to \mathrm{G} : (g, h) \mapsto gh, \qquad \mathrm{G} \to \mathrm{G} : g \mapsto g^{-1}$$

are holomorphic.

The Lie algebra $\mathfrak{g} := \mathrm{Lie}(\mathrm{G})$ of a complex Lie group G is equipped with a linear complex structure $\mathfrak{g} \to \mathfrak{g} : \zeta \mapsto \mathbf{i}\zeta$ that is preserved by the adjoint action of G, so the Lie bracket is complex bilinear. Conversely, if the Lie algebra is equipped with a linear complex structure $\mathbf{i}$ that is preserved by the adjoint action, then the formula $g^{-1}J(g)\widehat{g} := \mathbf{i}(g^{-1}\widehat{g})$ for $\widehat{g} \in T_g\mathrm{G}$ defines an integrable almost complex structure on G with respect to which the structure maps are holomorphic. Here integrability follows from the fact that the almost complex structure is preserved by the torsion-free connection $g^{-1}\nabla_t\widehat{g} = \frac{d}{dt}(g^{-1}\widehat{g}) + \frac{1}{2}[g^{-1}\dot{g}, g^{-1}\widehat{g}]$.

Theorem B.2 *Let* G *be a compact Lie group and let* G^c *be a complex Lie group with Lie algebras* $\mathfrak{g} := \mathrm{Lie}(\mathrm{G})$ *and* $\mathfrak{g}^c = \mathrm{Lie}(\mathrm{G}^c)$. *Let*

$$\iota : \mathrm{G} \to \mathrm{G}^c$$

be a Lie group homomorphism. Then the following are equivalent.

(i) *For every complex Lie group* H *and every Lie group homomorphism* $\rho : \mathrm{G} \to \mathrm{H}$ *there exists a unique holomorphic homomorphism* $\rho^c : \mathrm{G}^c \to \mathrm{H}$ *such that*

$$\rho^c \circ \iota = \rho.$$

© The Author(s), under exclusive license to Springer Nature Switzerland AG 2021
V. Georgoulas et al., *The Moment-Weight Inequality and the Hilbert–Mumford Criterion*, Lecture Notes in Mathematics 2297,
https://doi.org/10.1007/978-3-030-89300-2

(ii) *The homomorphism ι is injective, its image $\iota(\mathrm{G})$ is a maximal compact subgroup of G^c, the quotient $\mathrm{G}^c/\iota(\mathrm{G})$ is connected, and the derivative $d\iota(\mathbb{1}) : \mathfrak{g} \to \mathfrak{g}^c$ maps $\mathfrak{g}$ onto a totally real subspace of $\mathfrak{g}^c$.*

Proof See pages 164 and 167. □

A Lie group homomorphism $\iota : \mathrm{G} \to \mathrm{G}^c$ that satisfies the equivalent conditions of Theorem B.2 is called a **complexification** of G. By the universality property in part (i) of Theorem B.2, the complexification (G^c, ι) of a compact Lie group G is unique up to canonical isomorphism. A complex Lie group is called **reductive** iff it is the complexification of a compact Lie group.

Theorem B.3 *Every compact Lie group admits a complexification, unique up to canonical isomorphism.*

Proof See page 167. □

The archetypal example of a complexification is the inclusion of the unitary group $\mathrm{U}(n)$ into $\mathrm{GL}(n, \mathbb{C})$. Polar decomposition gives rise to a diffeomorphism

$$\phi : \mathrm{U}(n) \times \mathfrak{u}(n) \to \mathrm{GL}(n, \mathbb{C}), \qquad \phi(u, \eta) := \exp(\mathbf{i}\eta)u. \tag{B.1}$$

This example extends to every Lie subgroup of $\mathrm{U}(n)$.

Theorem B.4 *Let*

$$\mathrm{G} \subset \mathrm{U}(n)$$

be a Lie subgroup with Lie algebra

$$\mathfrak{g} := \mathrm{Lie}(\mathrm{G}) \subset \mathfrak{u}(n).$$

Then the set

$$\mathrm{G}^c := \{\exp(\mathbf{i}\eta)u \mid u \in \mathrm{G},\ \eta \in \mathfrak{g}\} \subset \mathrm{GL}(n, \mathbb{C})$$

is a complex Lie subgroup of $\mathrm{GL}(n, \mathbb{C})$ *and the inclusion of* G *into* G^c *satisfies condition (ii) in Theorem B.2. Moreover,* G^c/G *is diffeomorphic to* $\mathfrak{g}$.

Proof The proof has ten steps.

Step 1 G^c *is a closed submanifold of* $\mathrm{GL}(n, \mathbb{C})$.

This follows from the fact that (B.1) is a diffeomorphism.

Step 2 $\mathbb{1} \in \mathrm{G}^c$ and $T_{\mathbb{1}}\mathrm{G} = \mathfrak{g} \oplus \mathbf{i}\mathfrak{g} =: \mathfrak{g}^c$.

For $\xi, \eta \in \mathfrak{g}$ and $t \in \mathbb{R}$ define

$$\gamma(t) := \exp(\mathbf{i}t\eta)\exp(t\xi) \in \mathrm{G}^c.$$

Then

$$\dot{\gamma}(0) = \xi + \mathbf{i}\eta.$$

Hence

$$\mathfrak{g}^c \subset T_{\mathbb{1}}\mathrm{G}^c$$

and both spaces have the same dimension.

Step 3 $T_g\mathrm{G}^c = g\mathfrak{g}^c$ *for every* $g \in \mathrm{G}^c$.

Both spaces have the same dimension, so it suffices to prove that $T_g\mathrm{G}^c \subset g\mathfrak{g}^c$. Let ϕ be the diffeomorphism (B.1). Let $(u, \eta) \in \mathrm{G} \times \mathfrak{g}$ and

$$g := \phi(u, \eta) = \exp(\mathbf{i}\eta)u \in \mathrm{G}^c.$$

Then, for every $\widehat{u} \in T_u\mathrm{G}$, we have $d\phi(u, \eta)(\widehat{u}, 0) = \exp(\mathbf{i}\eta)u(u^{-1}\widehat{u}) \in g\mathfrak{g}^c$. Now let $\widehat{\eta} \in \mathfrak{g}$. We must prove that $d\phi(u, \eta)(0, \widehat{\eta}) \in g\mathfrak{g}^c$. To see this, define

$$\gamma(s, t) := \phi(u, t(\eta + s\widehat{\eta})) = \exp(\mathbf{i}t(\eta + s\widehat{\eta}))u$$

and

$$\xi(s, t) := \gamma(s, t)^{-1}\partial_s\gamma(s, t), \qquad \eta(s, t) := \gamma(s, t)^{-1}\partial_t\gamma(s, t)$$

for $s, t \in \mathbb{R}$. Then $\eta(s, t) = u^{-1}\mathbf{i}(\eta + s\widehat{\eta})u$ and

$$\partial_t\xi(s, t) = \partial_s\eta(s, t) + [\xi(s, t), \eta(s, t)], \qquad \xi(s, 0) = 0$$

for all $s, t \in \mathbb{R}$. Thus $\eta(s, t) \in \mathfrak{g}^c$ and hence $\xi(s, t) \in \mathfrak{g}^c$ for all s, t. In particular, we have $d\phi(u, \eta)(0, \widehat{\eta}) = \gamma(0, 1)\xi(0, 1) \in g\mathfrak{g}^c$ and this proves Step 3.

Step 4 *Let* $g \in \mathrm{GL}(n, \mathbb{C})$. *Then* $g \in \mathrm{G}^c$ *if and only if there exists a smooth curve* $\gamma : [0, 1] \to \mathrm{GL}(n, \mathbb{C})$ *such that*

$$\gamma(0) \in \mathrm{G}, \qquad \gamma(1) = g, \qquad \gamma(t)^{-1}\dot{\gamma}(t) \in \mathfrak{g}^c$$

for all t.

If $g = \exp(i\eta)u \in \mathrm{G}^c$, then $\gamma(t) := \exp(\mathbf{i}t\eta)u$ satisfies the requirements of Step 4. Conversely, let $\gamma : [0, 1] \to \mathrm{GL}(n, \mathbb{C})$ be a smooth curve with $\gamma(0) \in \mathrm{G}$, $\gamma(1) = g$, and $\gamma(t)^{-1}\dot{\gamma}(t) \in \mathfrak{g}^c$ for all t. Then the set $I := \{t \in [0, 1] \,|\, \gamma(t) \in \mathrm{G}^c\}$ is nonempty because $0 \in I$ and is closed because G^c is a closed subset of $\mathrm{GL}(n, \mathbb{C})$ by Step 1. To prove it is open, let $\eta(t) := \gamma(t)^{-1}\dot{\gamma}(t) \in \mathfrak{g}^c$ and define the vector field v_t on $\mathbb{C}^{n\times n}$ by $v_t(A) := A\eta(t)$. By Step 3, v_t is tangent to G^c. Hence every solution of the differential equation $\dot{A}(t) = A(t)\eta(t)$ that starts in G^c remains in G^c on a sufficiently small time interval. Hence I is open. Thus $I = [0, 1]$ and so $g = \gamma(1) \in G^c$.

Step 5 *If* $g \in G^c$ *and* $\zeta \in \mathfrak{g}^c$ *then* $g^{-1}\zeta g \in \mathfrak{g}^c$.

Choose $\gamma : [0, 1] \to G^c$ as in Step 4 with $\gamma(0) \in G$ and $\gamma(1) = g$ and define

$$\zeta(t) := \gamma(t)^{-1}\zeta\gamma(t), \qquad \zeta'(t) := \gamma(t)^{-1}\dot{\gamma}(t).$$

Then $\zeta'(t) \in \mathfrak{g}^c$ for all t by Step 3 and

$$\dot{\zeta}(t) + [\zeta'(t), \zeta(t)] = 0, \qquad \zeta(0) = \gamma(0)\zeta\gamma(0)^{-1} \in \mathfrak{g}^c.$$

Hence $\zeta(t) \in \mathfrak{g}^c$ for all t, and so $g^{-1}\zeta g = \zeta(1) \in \mathfrak{g}^c$.

Step 6 *If* $g \in G^c$ *and* $\zeta \in \mathfrak{g}^c$ *then* $g\zeta g^{-1} \in \mathfrak{g}^c$.

The linear map $\zeta \mapsto g^{-1}\zeta g$ maps $\mathfrak{g}^c$ to itself, by Step 5, and it is injective. Hence the map $\mathfrak{g}^c \to \mathfrak{g}^c : \zeta \mapsto g^{-1}\zeta g$ is bijective and this proves Step 6.

Step 7 *If* $g_0, g_1 \in G^c$ *then* $g_0g_1 \in G^c$.

Choose two curves $\gamma_0, \gamma_1 : [0, 1] \to G^c$ as in Step 4 with $\gamma_i(0) \in G$ and $\gamma_i(1) = g_i$. Then the curve $\gamma := \gamma_0\gamma_1 : [0, 1] \to \mathrm{GL}(n, \mathbb{C})$ satisfies

$$\gamma^{-1}\dot{\gamma} = \gamma_1^{-1}\dot{\gamma}_1 + \gamma_1^{-1}(\gamma_0^{-1}\dot{\gamma}_0)\gamma_1, \qquad \gamma(0) \in G.$$

By Step 5, $\gamma(t)^{-1}\dot{\gamma}(t) \in \mathfrak{g}^c$ for all t and hence, by Step 4, $g_0g_1 = \gamma(1) \in G^c$.

Step 8 *If* $g \in G^c$ *then* $g^{-1} \in G^c$.

Let $\gamma : [0, 1] \to G^c$ be as in Step 4 with $\gamma(0) \in G$ and $\gamma(1) = g$, and define the curve $\beta : [0, 1] \to \mathrm{GL}(n, \mathbb{C})$ by

$$\beta(t) := \gamma(t)^{-1}$$

for $0 \le t \le 1$. Then $\beta(0) \in G$ and

$$\beta^{-1}\dot{\beta} = \gamma\frac{d}{dt}\gamma^{-1} = -\dot{\gamma}\gamma^{-1} = \gamma(-\gamma^{-1}\dot{\gamma})\gamma^{-1}.$$

Hence $\beta(t)^{-1}\dot{\beta}(t) \in \mathfrak{g}^c$ for all t by Step 6, and so $g^{-1} = \beta(1) \in G^c$ by Step 4.

Step 9 G^c *is a complex Lie subgroup of* $\mathrm{GL}(n, \mathbb{C})$.

By Step 3 G^c is a complex submanifold of $\mathrm{GL}(n, \mathbb{C})$, and by Steps 7 and 8 it is a subgroup of $\mathrm{GL}(n, \mathbb{C})$.

Step 10 G *is a maximal compact subgroup of* G^c *and* G^c/G *is diffeomorphic to* $\mathfrak{g}$.

That G^c/G is diffeomorphic to $\mathfrak{g}$ follows directly from the definition. Let $H \subset G^c$ be a subgroup such that $G \subsetneq H$. Choose an element $h \in H \setminus G$. Since $H \subset G^c$, there is a pair $(u, \eta) \in G \times \mathfrak{g}$ such that

$$h = \exp(\mathbf{i}\eta)u.$$

Since $G \subset H$ and H is a subgroup of G^c we have

$$P := \exp(\mathbf{i}\eta) \in H.$$

The matrix P is Hermitian and positive definite. Since $h \notin G$ we also have $P \notin G$. But this implies $\eta \neq 0$ and so at least one eigenvalue of P is not equal to 1. Hence the sequence

$$P^k = \exp(\mathbf{i}k\eta) \in H, \qquad k = 1, 2, 3, \dots$$

has no subsequence that converges to an element of $\mathrm{GL}(n, \mathbb{C})$. Thus H is not compact and this proves Theorem B.4. □

Remark B.5 Here is a sketch of a proof of Theorem B.4 in the intrinsic setting. (For a more detailed proof see [44].) If G is a Lie group, then the formula

$$A(g)\widehat{g} := g^{-1}\widehat{g} \tag{B.2}$$

for $g \in G$ and $\widehat{g} \in T_g G$ defines a flat connection $A \in \Omega^1(G, \mathfrak{g})$ such that

(a) for every $g \in G$ the linear map $A(g) : T_g G \to \mathfrak{g}$ is bijective,
(b) for every smooth path

$$\zeta : \mathbb{R} \to \mathfrak{g}$$

and every element $g \in G$ the differential equation

$$A(\gamma(t))\dot{\gamma}(t) = \zeta(t), \qquad \gamma(0) = g$$

has a solution $\gamma : \mathbb{R} \to G$ (on all of $\mathbb{R}$),
(c) the holonomy is trivial, i.e. if $\gamma : [0, 1] \to G$ is a smooth curve with $\gamma(0)=\gamma(1)$ then every solution $\zeta : [0, 1] \to \mathfrak{g}$ of the differential equation

$$\dot{\zeta} + [A(\gamma)\dot{\gamma}, \zeta] = 0$$

satisfies $\zeta(0) = \zeta(1)$.

Conversely, if $A \in \Omega^1(G, \mathfrak{g})$ is a Lie algebra valued 1-form on a connected manifold G that satisfies (a), (b), and (c), and $\mathbb{1}$ is any element of G, then G has the unique structure of a Lie group with unit $\mathbb{1}$ such that A is given by (B.2).

Now let G be a compact Lie group with Lie algebra $\mathfrak{g}$ and define $\mathfrak{g}^c := \mathfrak{g} \oplus \mathbf{i}\mathfrak{g}$. Then there exists a unique flat connection $B \in \Omega^1(\mathfrak{g}, \mathfrak{g}^c)$ such that

$$[\eta, \widehat{\eta}] = 0 \qquad \Longrightarrow \qquad B(\eta)\widehat{\eta} = \mathbf{i}\widehat{\eta} \tag{B.3}$$

for all $\eta, \widehat{\eta} \in \mathfrak{g}$. (Define $B(\eta)\widehat{\eta} := \zeta(1)$, where $\zeta : [0,1] \to \mathfrak{g}^c$ is the unique solution of $\dot{\zeta} + [\mathbf{i}\eta, \zeta] = \mathbf{i}\widehat{\eta}$ with $\zeta(0) = 0$.) Now define

$$A(u,\eta)(\widehat{u},\widehat{\eta}) := u^{-1}\widehat{u} + u^{-1}(B(\eta)\widehat{\eta})u \tag{B.4}$$

for $u \in \mathrm{G}$, $\widehat{u} \in T_u\mathrm{G}$, and $\eta, \widehat{\eta} \in \mathfrak{g}$. Then

$$A \in \Omega^1(\mathrm{G} \times \mathfrak{g}, \mathfrak{g}^c)$$

is a flat connection satisfying (a), (b), (c) and hence gives rise to a unique group structure on

$$\mathrm{G}^c := \mathrm{G} \times \mathfrak{g}$$

such that the map $\iota : \mathrm{G} \to \mathrm{G}^c$, defined by $\iota(u) := (u, 0)$, is a group homomorphism. Moreover, the homorphism

$$\iota : \mathrm{G} \to \mathrm{G}^c$$

satisfies condition (ii) in Theorem B.2 and this gives rise to a proof of Theorem B.4 in the intrinsic setting.

Remark B.6 If G^c satisfies condition (ii) in Theorem B.2, then the homogeneous space G^c/G is simply connected. The proof requires the following steps.

Step 1 *If* $\eta \in \mathfrak{g}$ *and* $\exp(\mathbf{i}\eta) \in \mathrm{G}$ *then* $[\xi, \eta] = 0$ for all $\xi \in \mathfrak{g}$.

Step 2 *If* $\eta \in \mathfrak{g}$ *and* $\exp(\mathbf{i}\eta) \in \mathrm{G}$ *then* $\eta = 0$.

Step 3 G^c/G *is simply connected.*

The proof of Step 1 uses the fact that G^c/G has nonpositive sectional curvature (see Appendix C). The proof of Step 2 uses Step 1 and the fact that G is a maximal compact subgroup of G^c (which by the Cartan–Iwasawa–Malcev Theorem in [50, Thm 14.1.3] implies that the intersection of G with the identity component G_0^c of G^c is a maximal compact subgroup of G_0^c). The proof of Step 3 uses Step 2 and the existence of a nonconstant closed geodesic in each nontrivial homotopy class. It follows from Step 3 and Hadamard's theorem (e.g. [70, Theorem 6.5.2]) that the map

$$\mathrm{G} \times \mathfrak{g} \to \mathrm{G}^c : (u, \eta) \mapsto \phi(u,\eta) := \exp(\mathbf{i}\eta)u$$

is a diffeomorphism. This is the **Cartan Decomposition Theorem**.

Proof of Theorem B.2 "(ii) $\Longrightarrow$ (i)" If G^c satisfies condition (ii) in Theorem B.2, then the homogeneous space G^c/G is connected and simply connected (see Remark B.6). Hence the following holds.

(I) *For every* $g \in \mathrm{G}^c$ *there exists a smooth path* $\gamma : [0, 1] \to \mathrm{G}^c$ *such that* $\gamma(0) \in \mathrm{G}$ *and* $\gamma(1) = g$.

(II) *Any two paths* $\gamma_0, \gamma_1 : [0, 1] \to \mathrm{G}^c$ *as in (I) can be joined by a smooth homotopy* $\{\gamma_s\}_{0 \le s \le 1}$ *satisfying* $\gamma_s(0) \in \mathrm{G}$ *and* $\gamma_s(1) = g$ for all $s \in [0, 1]$.

Now let H be a complex Lie group, let $\mathfrak{h} := \mathrm{Lie}(\mathrm{H})$ be its Lie algebra, let

$$\rho : \mathrm{G} \to \mathrm{H}$$

be a Lie group homomorphism, let

$$\Phi := d\rho(1) : \mathfrak{g} \to \mathfrak{h}$$

be the induced Lie algebra homomorphism, and denote its complexification by

$$\Phi^c : \mathfrak{g}^c \to \mathfrak{h}.$$

We define the extended map $\rho^c : \mathrm{G}^c \to \mathrm{H}$ as follows. Given an element $g \in \mathrm{G}^c$ choose γ as in (I), let

$$\beta : [0, 1] \to \mathrm{H}$$

be the unique solution of the differential equation

$$\beta^{-1}\dot{\beta} = \Phi^c(\gamma^{-1}\dot{\gamma}), \qquad \beta(0) = \rho(\gamma(0)), \tag{B.5}$$

and define

$$\rho^c(g) := \beta(1).$$

We prove that ρ^c is well defined, i.e. that $\beta(1)$ does not depend on the choice of the path γ. By (II) any two paths γ_0 and γ_1 with $\gamma_0(0), \gamma_1(0) \in \mathrm{G}$ and $\gamma_0(1) = \gamma_1(1) = g$ can be joined by a smooth homotopy $[0, 1]^2 \to \mathrm{G}^c : (s, t) \mapsto \gamma_s(t) = \gamma(s, t)$ such that $\gamma_s(0) \in \mathrm{G}$ and $\gamma_s(1) = g$ for all s. Define $\beta : [0, 1]^2 \to \mathrm{H}$ by

$$\beta^{-1}\partial_t\beta = \Phi^c(\gamma^{-1}\partial_t\gamma), \qquad \beta(s, 0) = \rho(\gamma(s, 0)).$$

We claim that

$$\beta^{-1}\partial_s\beta = \Phi^c(\gamma^{-1}\partial_s\gamma). \tag{B.6}$$

To see this, abbreviate

$$\zeta_s := \gamma^{-1}\partial_s\gamma, \qquad \zeta_t := \gamma^{-1}\partial_t\gamma, \qquad \eta_s := \beta^{-1}\partial_s\beta, \qquad \eta_t := \beta^{-1}\partial_t\beta.$$

Then $\eta_t = \Phi^c(\zeta_t)$ by definition of β and

$$\partial_t \eta_s = \partial_s \eta_t + [\eta_s, \eta_t], \qquad \partial_t \Phi^c(\zeta_s) = \partial_s \Phi^c(\zeta_t) + [\Phi^c(\zeta_s), \Phi^c(\zeta_t)].$$

Moreover, when $t = 0$ we have

$$d\rho(\gamma)\partial_s\gamma = d\rho(\gamma)\gamma\zeta_s = \rho(\gamma)\Phi(\zeta_s)$$

and hence

$$\eta_s(s, 0) = \beta(s, 0)^{-1}\partial_s\beta(s, 0) = \Phi(\gamma(s, 0)^{-1}\partial_s\gamma(s, 0)) = \Phi(\zeta_s(s, 0)).$$

Hence both curves $t \mapsto \eta_s(s, t)$ and $t \mapsto \Phi^c(\zeta_s(s, t))$ satisfy the same initial value problem and so they agree. This proves (B.6). Hence

$$\eta_s(s, 1) = \Phi^c(\zeta_s(s, 1)) = 0$$

and therefore $\partial_s\beta(s, 1) = 0$. This shows that ρ^c is well defined.

We prove that, for $g \in \mathrm{G}^c$ and $\zeta \in \mathfrak{g}^c$, we have

$$\Phi^c(g^{-1}\zeta g) = \rho^c(g)^{-1}\Phi^c(\zeta)\rho^c(g). \tag{B.7}$$

Choose γ and β as in the definition of $\rho^c(g)$, so that $\beta^{-1}\dot{\beta} = \Phi^c(\gamma^{-1}\dot{\gamma})$, and define

$$\zeta(t) := \gamma(t)^{-1}\zeta\gamma(t), \qquad \eta(t) := \beta(t)^{-1}\Phi^c(\zeta)\beta(t).$$

Then $\eta(t)$ and $\Phi^c(\zeta(t))$ satsify the same differential equation

$$\dot{\eta}(t) + [\beta(t)^{-1}\dot{\beta}(t), \eta(t)] = 0$$

and the same initial condition

$$\eta(0) = \rho(\gamma(0))^{-1}\Phi^c(\zeta)\rho(\gamma(0)) = \Phi^c(\zeta(0)).$$

Hence they have the same endpoints and this proves Eq. (B.7).

We prove that ρ^c is a group homomorphism. Let $g_0, g_1 \in \mathrm{G}^c$ and choose γ_i and β_i as in the definition of $\rho^c(g_i)$ for $i = 0, 1$. Then $\rho^c(\gamma_i(t)) = \beta_i(t)$ for $0 \le t \le 1$ and $i = 0, 1$. Define

$$\gamma := \gamma_0\gamma_1, \qquad \beta := \beta_0\beta_1.$$

Then, by (B.7), we have

$$
\begin{aligned}
\beta^{-1}\dot{\beta} &= \beta_1^{-1}\dot{\beta}_1 + \beta_1^{-1}\beta_0^{-1}\dot{\beta}_0\beta_1 \\
&= \Phi^c(\gamma_1^{-1}\dot{\gamma}_1) + \rho^c(\gamma_1)^{-1}\Phi^c(\gamma_0^{-1}\dot{\gamma}_0)\rho^c(\gamma_1) \\
&= \Phi^c(\gamma_1^{-1}\dot{\gamma}_1 + \gamma_1^{-1}\gamma_0^{-1}\dot{\gamma}_0\gamma_1) \\
&= \Phi^c(\gamma^{-1}\dot{\gamma}).
\end{aligned}
$$

Hence $\rho^c(g_0g_1) = \beta(1) = \beta_0(1)\beta_1(1) = \rho^c(g_0)\rho^c(g_1)$ and so ρ^c is a group homomorphism.

We prove that ρ^c is smooth. Consider the commutative diagram

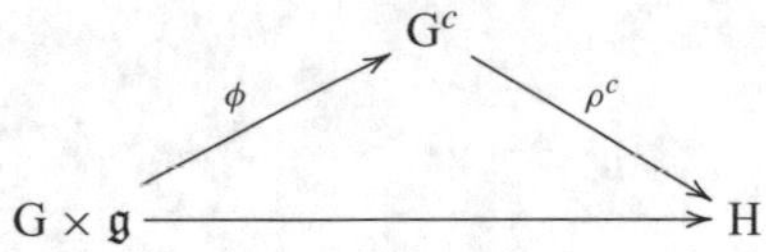

Here $\phi : \mathrm{G} \times \mathfrak{g} \to \mathrm{G}^c$ is the diffeomorphism in Theorem B.4, i.e. $\phi(u, \eta) = \exp(\mathbf{i}\eta)u$ for $u \in \mathrm{G}$ and $\eta \in \mathfrak{g}$. In the intrinsic setting the Cartan Decomposition Theorem asserts that this map is a diffeomorphism under the assumption that G^c satisfies condition (ii) in Theorem B.2 (see Remark B.6). The map $\mathrm{G} \times \mathfrak{g} \to \mathrm{H}$ is given by $(u, \eta) \mapsto \exp(\mathbf{i}\Phi^c(\eta))\rho(u)$ and hence is smooth. That the differential of ρ^c at $\mathbb{1}$ is given by Φ^c follows also from this diagram. This proves existence, and uniqueness is obvious. Thus we have proved that (ii) implies (i) in Theorem B.2. □

Proof of Theorem B.3 By Theorem B.4 (respectively Remark B.5 in the intrinsic setting), there exists an embedding $\iota : \mathrm{G} \to \mathrm{G}^c$ into a complex Lie group (diffeomorphic to $\mathrm{G} \times \mathfrak{g}$) that satisfies condition (ii) in Theorem B.2. Since (ii) implies (i) in Theorem B.2, the embedding $\iota : \mathrm{G} \to \mathrm{G}^c$ satisfies both (i) and (ii) in Theorem B.2 and hence is a complexification. Moreover, any two embeddings of G into a complex Lie group that satisfy (i) in Theorem B.2 are naturally isomorphic. This proves Theorem B.3. □

Proof of Theorem B.2 "(i) $\Longrightarrow$ (ii)" Let $\iota : \mathrm{G} \to \mathrm{G}^c$ be an embedding into a complex Lie group that satisfies (i). By Theorem B.3 there exists an embedding $\widetilde{\iota} : \mathrm{G} \to \widetilde{\mathrm{G}}^c$ into a complex Lie group that satisfies both (i) and (ii). Since both embeddings satisfy (i), there exists a unique holomorphic Lie group isomorphism $\phi : \mathrm{G}^c \to \widetilde{\mathrm{G}}^c$ such that $\phi \circ \iota = \widetilde{\iota}$. Since the embedding $\widetilde{\iota}$ satisfies (ii), so does ι. This proves Theorem B.2. □

Appendix C
The Homogeneous Space $M = G^c/G$

Let $G \subset U(n)$ be a compact Lie group and denote by $G^c \subset GL(n, \mathbb{C})$ the complexified group. Then the homogeneous space

$$M := G^c/G := \left\{\pi(g) \mid g \in G^c\right\},$$

is a connected, simply connected, complete Riemannian manifold with nonpositive sectional curvature. The purpose of the present appendix to explain this basic fact. Denote by $\pi : G^c \to M$ the canonical projection, given by $\pi(g) := gG$ for $g \in G^c$.

Theorem C.1 *Choose an invariant inner product on* $\mathfrak{g}$ *and define a Riemannian metric on* M *by*

$$\langle v_1, v_2\rangle_p := \langle \eta_1, \eta_2\rangle, \qquad p = \pi(g), \qquad v_i = d\pi(g)g\mathbf{i}\eta_i, \tag{C.1}$$

for $g \in G^c$ *and* $\eta_1, \eta_2 \in \mathfrak{g}$.

(i) *Let* $g : \mathbb{R} \to G^c$ *and* $\eta : \mathbb{R} \to \mathfrak{g}$ *be smooth curves. Then the covariant derivative of the vector field* $X := d\pi(g)g\mathbf{i}\eta \in \mathrm{Vect}(\gamma)$ *along the curve* $\gamma := \pi \circ g : \mathbb{R} \to M$ *is given by*

$$\nabla X = d\pi(g)g\mathbf{i}\Big(\dot\eta + [\mathrm{Re}(g^{-1}\dot g), \eta]\Big). \tag{C.2}$$

(ii) *The geodesics on* M *have the form*

$$\gamma(t) = \pi(g\exp(\mathbf{i}t\eta))$$

for $g \in G^c$ *and* $\eta \in \mathfrak{g}$.

© The Author(s), under exclusive license to Springer Nature Switzerland AG 2021
V. Georgoulas et al., *The Moment-Weight Inequality and the Hilbert–Mumford Criterion*, Lecture Notes in Mathematics 2297,
https://doi.org/10.1007/978-3-030-89300-2

(iii) *The Riemann curvature tensor on* G^c/G *is given by*

$$R_p(v_1, v_1)v_3 = d\pi(g)g\mathbf{i}[[\eta_1, \eta_2], \eta_3], \qquad p = \pi(g), \qquad v_i = d\pi(g)g\mathbf{i}\eta_i.$$

for $g \in \mathrm{G}^c$ *and* $\eta_i \in \mathfrak{g}$.

(iv) M *is a complete, connected, simply connected Riemannian manifold of non-positive sectional curvature.*

Proof The projection $\pi : \mathrm{G}^c \to M$ is a principal G-bundle. The formula

$$A_g(\widehat{g}) := \mathrm{Re}(g^{-1}\widehat{g})$$

defines a connection 1-form $A \in \Omega^1(\mathrm{G}^c, \mathfrak{g})$. The map

$$\mathrm{G}^c \times \mathfrak{g} \to TM : (g, \eta) \mapsto d\pi(g)g\mathbf{i}\eta.$$

descends to a vector bundle isomorphism from the associated bundle $\mathrm{G}^c \times_{\mathrm{ad}} \mathfrak{g}$ to the tangent bundle of M. Thus A induces a connection on TM and this connection is given by (C.2). It is a Riemannian connection because the inner product on the Lie algebra $\mathfrak{g}$ is invariant under conjugation. We prove that the connection (C.2) is torsion-free. Denote by s and t the standard coordinates on $\mathbb{R}^2$. Choose a smooth map $g : \mathbb{R}^2 \to \mathrm{G}^c$ and denote $\gamma := \pi \circ g$. Then

$$\begin{aligned}
\nabla_s \partial_t \gamma &= d\pi(g)g\mathbf{i}\Big(\partial_s \mathrm{Im}(g^{-1}\partial_t g) + [\mathrm{Re}(g^{-1}\partial_s g), \mathrm{Im}(g^{-1}\partial_t g)]\Big) \\
&= d\pi(g)g\mathbf{i}\Big(\partial_t \mathrm{Im}(g^{-1}\partial_s g) + [\mathrm{Re}(g^{-1}\partial_t g), \mathrm{Im}(g^{-1}\partial_s g)]\Big) \\
&= \nabla_t \partial_s \gamma
\end{aligned}$$

Here the second equation follows from the identity

$$\partial_s(g^{-1}\partial_t g) - \partial_t(g^{-1}\partial_s g) + [g^{-1}\partial_s g, g^{-1}\partial_t g] = 0. \tag{C.3}$$

This proves (i).

We prove part (ii). A smooth curve

$$\gamma(t) = \pi(g(t))$$

is a geodesic if and only if $\nabla\dot{\gamma} \equiv 0$. By (i) this is equivalent to the differential equation

$$\partial_t \mathrm{Im}(g^{-1}\dot{g}) + [\mathrm{Re}(g^{-1}\dot{g}), \mathrm{Im}(g^{-1}\dot{g})] = 0.$$

A curve $g : \mathbb{R} \to \mathrm{G}^c$ satisfies this equation if and only if it has the form

$$g(t) = g_0 \exp(\mathbf{i}t\eta)u(t)$$

for some $g_0 \in \mathrm{G}^c$, $\eta \in \mathfrak{g}$, and $u : \mathbb{R} \to \mathrm{G}$. This proves (ii).

We prove part (iii). Choose maps $g : \mathbb{R}^2 \to \mathrm{G}^c$ and $\eta : \mathbb{R}^2 \to \mathfrak{g}$ and define

$$\zeta_s := g^{-1}\partial_s g, \qquad \zeta_t := g^{-1}\partial_t g, \qquad \partial_s\zeta_t - \partial_t\zeta_s + [\zeta_s, \zeta_t] = 0. \tag{C.4}$$

Here the third equation follows from (C.3). Now define

$$\gamma := \pi \circ g, \qquad Z_s := \partial_s\gamma = d\pi(g)g\zeta_s, \qquad Z_t := \partial_t\gamma = d\pi(g)g\zeta_t,$$

and

$$Y := d\pi(g)g\mathbf{i}\eta.$$

Then, by part (i), we have

$$\nabla_s Y = d\pi(g)g\mathbf{i}\Big(\partial_s\eta + [\mathrm{Re}(\zeta_s), \eta]\Big),$$
$$\nabla_t Y = d\pi(g)g\mathbf{i}\Big(\partial_t\eta + [\mathrm{Re}(\zeta_t), \eta]\Big).$$

Hence $R(Z_s, Z_t)Y = \nabla_s\nabla_t Y - \nabla_t\nabla_s Y = d\pi(g)g\mathbf{i}\widetilde{\eta}$, where

$$\begin{aligned}
\widetilde{\eta} &= \partial_s\Big(\partial_t\eta + [\mathrm{Re}(\zeta_t), \eta]\Big) + \Big[\mathrm{Re}(\zeta_s), \Big(\partial_t\eta + [\mathrm{Re}(\zeta_t), \eta]\Big)\Big] \\
&\quad - \partial_t\Big(\partial_s\eta + [\mathrm{Re}(\zeta_s), \eta]\Big) - \Big[\mathrm{Re}(\zeta_t), \Big(\partial_s\eta + [\mathrm{Re}(\zeta_s), \eta]\Big)\Big] \\
&= [\mathrm{Re}(\partial_s\zeta_t), \eta] + [\mathrm{Re}(\zeta_s), [\mathrm{Re}(\zeta_t), \eta]] \\
&\quad - [\mathrm{Re}(\partial_t\zeta_s), \eta] - [\mathrm{Re}(\zeta_t), [\mathrm{Re}(\zeta_s), \eta]] \\
&= \Big[\mathrm{Re}(\partial_s\zeta_t) - \mathrm{Re}(\partial_t\zeta_s) + [\mathrm{Re}(\zeta_s), \mathrm{Re}(\zeta_t)], \eta\Big] \\
&= [[\mathrm{Im}(\zeta_s), \mathrm{Im}(\zeta_t)], \eta].
\end{aligned}$$

Here the last equality follows from (C.4). This proves (iii). By (iii), we have

$$\langle R(Z_s, Z_t)Z_t, Z_s\rangle = -\,|[\mathrm{Im}(\zeta_s), \mathrm{Im}(\zeta_t)]|^2 \le 0.$$

This proves Theorem C.1. □

Lemma C.2 *If* $\xi_0, \xi_1, \eta \in \mathfrak{g}$ *satisfy*

$$\exp(-\mathbf{i}\xi_1)\exp(\mathbf{i}\xi_0)\exp(\mathbf{i}\eta) \in \mathrm{G},$$

then $|\xi_0 - \xi_1| \leq |\eta|$.

Proof Define $g_0 := \exp(\mathbf{i}\xi_0)$ and $g_1 := \exp(\mathbf{i}\xi_1)$. Then the unique geodesic in G^c/G connecting $\pi(g_0)$ to $\pi(g_1)$ is given by $\gamma(t) := \pi(g_0\exp(\mathbf{i}t\eta))$ for $0 \leq t \leq 1$. This implies $d(\pi(g_0), \pi(g_1)) = |\eta|$. Therefore it follows from Lemma A.4 with $M = \mathrm{G}^c/\mathrm{G}$, $p = \pi(\mathbb{1})$, $v_0 = d\pi(\mathbb{1})\mathbf{i}\xi_0$, $v_1 = d\pi(\mathbb{1})\mathbf{i}\xi_1$, $\exp_p(v_0) = \pi(g_0)$, and $\exp_p(v_1) = \pi(g_1)$ that $|\xi_0 - \xi_1| \leq |\eta|$. This proves Lemma C.2. □

Lemma C.3 *Every compact subgroup of* G^c *is conjugate to a subgroup of* G.

Proof Let $\mathrm{K} \subset \mathrm{G}^c$ be a compact subgroup. Then K acts on G^c/G by isometries via $k \cdot \pi(g) := \pi(kg)$ for $k \in \mathrm{K}$ and $g \in \mathrm{G}^c$. By Theorem A.5 the action of K on G^c/G has a fixed point $\pi(g) \in \mathrm{G}^c/\mathrm{G}$. Hence $\pi(kg) = \pi(g)$ and hence $g^{-1}kg \in \mathrm{G}$ for every $k \in \mathrm{K}$. This proves Lemma C.3. □

Lemma C.4 *Let* $\zeta \in \mathfrak{g}^c$. *Then the following are equivalent.*

(i) ζ *is semi-simple and has imaginary eigenvalues.*
(ii) *There is a* $g \in \mathrm{G}^c$ *such that* $g^{-1}\zeta g \in \mathfrak{g}$.

Proof Assume (i). Then the set $\mathrm{T} := \overline{\{\exp(t\zeta)\,|\,t \in \mathbb{R}\}} \subset \mathrm{G}^c$ is a (compact) torus. By Lemma C.3 there exists an element $g \in \mathrm{G}^c$ such that $g^{-1}\mathrm{T}g \subset \mathrm{G}$. This implies $g^{-1}\zeta g = \frac{d}{dt}|_{t=o}g^{-1}\exp(t\zeta)g \in \mathfrak{g}$. That (ii) implies (i) is obvious. □

Appendix D
Toral Generators

This appendix introduces toral generators and Mumford's equivalence relation. Let

$$\mathrm{G} \subset \mathrm{U}(n)$$

be a compact Lie group with the complexification

$$\mathrm{G}^c \subset \mathrm{GL}(n, \mathbb{C})$$

and denote their Lie algebras by

$$\mathfrak{g} := \mathrm{Lie}(\mathrm{G}) \subset \mathfrak{u}(n), \qquad \mathfrak{g}^c := \mathrm{Lie}(\mathrm{G}^c) = \mathfrak{g} + \mathbf{i}\mathfrak{g} \subset \mathfrak{gl}(n, \mathbb{C}).$$

Definition D.1 A nonzero element $\zeta \in \mathfrak{g}^c$ is called a **toral generator** iff it is semi-simple and has purely imaginary eigenvalues. This means that the subset

$$T_\zeta := \overline{\{\exp(t\zeta) \mid t \in \mathbb{R}\}}$$

is a torus in G^c. By Lemma C.4 the set of toral generators is

$$\mathscr{T}^c := \mathrm{ad}(\mathrm{G}^c)(\mathfrak{g} \setminus \{0\}).$$

Throughout we use the notation

$$\begin{aligned} \Lambda &:= \{\xi \in \mathfrak{g} \setminus \{0\} \mid \exp(\xi) = 1\}, \\ \Lambda^c &:= \left\{\zeta \in \mathfrak{g}^c \setminus \{0\} \mid \exp(\zeta) = 1\right\}. \end{aligned} \tag{D.1}$$

© The Author(s), under exclusive license to Springer Nature Switzerland AG 2021
V. Georgoulas et al., *The Moment-Weight Inequality and the Hilbert–Mumford Criterion*, Lecture Notes in Mathematics 2297,
https://doi.org/10.1007/978-3-030-89300-2

Thus

$$\Lambda \subset \Lambda^c \subset \mathscr{T}^c.$$

The elements of Λ^c are in one-to-one correspondence with nontrivial one-parameter subgroups $\mathbb{C}^* \to \mathrm{G}^c$. The set $\Lambda \cup \{0\}$ intersects the Lie algebra $\mathfrak{t} \subset \mathfrak{g}$ of any maximal torus $\mathrm{T} \subset \mathrm{G}$ in a spanning lattice and every element of Λ^c is conjugate to an element of $\Lambda \cap \mathfrak{t}$ (see Lemma C.4).

Lemma D.2 (Parabolic Subgroups) *For $\zeta \in \mathscr{T}^c$ the set*

$$\mathrm{P}(\zeta) := \left\{ p \in \mathrm{G}^c \mid \textit{the limit } \lim_{t\to\infty} \exp(\mathbf{i}t\zeta)p\exp(-\mathbf{i}t\zeta) \textit{ exists in } \mathrm{G}^c \right\} \tag{D.2}$$

is a Lie subgroup of G^c *with Lie algebra*

$$\mathfrak{p}(\zeta) := \left\{ \rho \in \mathfrak{g}^c \mid \textit{the limit } \lim_{t\to\infty} \exp(\mathbf{i}t\zeta)\rho\exp(-\mathbf{i}t\zeta) \textit{ exists in } \mathfrak{g}^c \right\}. \tag{D.3}$$

Proof Let $\zeta \in \mathscr{T}^c$. Then the matrix $\mathbf{i}\zeta \in \mathbb{C}^{n\times n}$ is semi-simple and has real eigenvalues, denoted by

$$\lambda_1 < \lambda_2 < \cdots < \lambda_k.$$

Denote the eigenspace of λ_j by V_j so we have an eigenspace decomposition

$$\mathbb{C}^n = V_1 \oplus V_2 \oplus \cdots \oplus V_k.$$

Write a matrix $\rho \in \mathfrak{g}^c \subset \mathfrak{gl}(n, \mathbb{C})$ in the form

$$\rho = \begin{pmatrix} \rho_{11} & \rho_{12} & \cdots & \rho_{1k} \\ \rho_{21} & \rho_{22} & \cdots & \rho_{2k} \\ \vdots & \vdots & \ddots & \vdots \\ \rho_{k1} & \rho_{k2} & \cdots & \rho_{kk} \end{pmatrix}, \qquad \rho_{ij} \in \mathrm{Hom}(V_j, V_i).$$

Then

$$\exp(\mathbf{i}t\zeta)\rho\exp(-\mathbf{i}t\zeta) = \begin{pmatrix} \rho_{11} & e^{(\lambda_1-\lambda_2)t}\rho_{12} & \cdots & e^{(\lambda_1-\lambda_k)t}\rho_{1k} \\ e^{(\lambda_2-\lambda_1)t}\rho_{21} & \rho_{22} & \cdots & e^{(\lambda_2-\lambda_k)t}\rho_{2k} \\ \vdots & \vdots & \ddots & \vdots \\ e^{(\lambda_k-\lambda_1)t}\rho_{k1} & e^{(\lambda_k-\lambda_2)t}\rho_{k2} & \cdots & \rho_{kk} \end{pmatrix}.$$

Thus $\rho \in \mathfrak{p}(\zeta)$ if and only if $\rho \in \mathfrak{g}^c$ and $\rho_{ij} = 0$ for $i > j$. Likewise, $g \in \mathrm{P}(\zeta)$ if and only if $g \in \mathrm{G}^c$ and $g_{ij} = 0$ for $i > j$. Hence $\mathrm{P}(\zeta)$ is a closed subset of G^c. Since every closed subgroup of a Lie group is a Lie subgroup, this proves Lemma D.2. □

The proof of Lemma D.2 shows that $\mathrm{P}(\zeta)$ is what is called in the theory of algebraic groups a **parabolic subgroup** of G^c (upper triangular matrices). It also shows that, for $\zeta = \xi \in \mathfrak{g}$, its intersection with G is the centralizer

$$\mathrm{P}(\xi) \cap \mathrm{G} = \mathrm{C}(\xi) := \left\{ u \in \mathrm{G} \mid u\xi u^{-1} = \xi \right\}.$$

In this case there is a G-equivariant isomorphism

$$\mathrm{G}^c/\mathrm{P}(\xi) \cong \mathrm{G}/\mathrm{C}(\xi). \tag{D.4}$$

For a generic element $\xi \in \mathfrak{g}$, the identity component

$$\mathrm{B} \subset \mathrm{P}(\xi)$$

is a **Borel subgroup** (a maximal parabolic subgroup) of G^c, the intersection

$$\mathrm{T} := \mathrm{B} \cap \mathrm{G}$$

is a maximal torus and is the identity component of the centralizer $C(\xi)$, and Eq. (D.4) reads

$$\mathrm{G}^c/\mathrm{B} \cong \mathrm{G}/\mathrm{T}.$$

In general, Eq. (D.4) can be restated as follows.

Theorem D.3 *For every $\xi \in \mathfrak{g} \setminus \{0\}$ and every $g \in \mathrm{G}^c$ there exists an element $p \in \mathrm{P}(\xi)$ such that $p^{-1}g \in \mathrm{G}$.*

Proof In Appendix E we give a proof of Theorem D.3 which does not rely on the structure theory of Lie groups. □

For the space $\Lambda^c \subset \mathscr{T}^c$ of generators of one-parameter subgroups of G^c the following obervations are essentially contained in [63, Chapter 2].

Theorem D.4 (Mumford) *Define a relation on $\mathscr{T}^c$ by*

$$\zeta \sim \zeta' \quad \stackrel{\text{def}}{\Longleftrightarrow} \quad \exists\, p \in \mathrm{P}(\zeta) \text{ such that } p\zeta p^{-1} = \zeta'. \tag{D.5}$$

The formula (D.5) *defines an equivalence relation on $\mathscr{T}^c$, invariant under conjugation, and every equivalence class contains a unique element of* $\mathfrak{g}$.

Proof The group G^c acts on the space

$$\Gamma^c := \bigsqcup_{\zeta \in \mathscr{T}^c} \mathrm{P}(\zeta)$$

by the diagonal adjoint action. This determines a groupoid. The elements of $\mathscr{T}^c$ are the objects of the groupoid. A pair $(\zeta, p) \in \Gamma^c$ is a morphism from ζ to $p\zeta p^{-1}$. The inverse map is given by $(\zeta, p) \mapsto (p\zeta p^{-1}, p^{-1})$ and the composition map sends a composable pair of pairs consisting of (ζ, p) and (ζ', p') with $\zeta' = p\zeta p^{-1}$ to the pair $(\zeta, p'p)$ with

$$p'p \in \mathrm{P}(p\zeta p^{-1})p = p\mathrm{P}(\zeta) = \mathrm{P}(\zeta).$$

This shows that (D.5) is an equivalence relation.

We prove that the equivalence relation (D.5) is invariant under conjugation. Choose equivalent elements $\zeta, \zeta' \in \mathscr{T}^c$ and let $g \in \mathrm{G}^c$. Then there exists an element $p \in \mathrm{P}(\zeta)$ such that $p\zeta p^{-1} = \zeta'$. Hence $gpg^{-1} \in \mathrm{P}(g\zeta g^{-1})$ and

$$(gpg^{-1})(g\zeta g^{-1})(gpg^{-1})^{-1} = g\zeta' g^{-1},$$

so $g\zeta g^{-1}$ is equivalent to $g\zeta' g^{-1}$.

Now let $\zeta \in \mathscr{T}^c$. By Lemma C.4, there is a $g \in \mathrm{G}^c$ such that

$$\xi := g\zeta g^{-1} \in \mathfrak{g}.$$

By Theorem D.3, there is a $q \in \mathrm{P}(\xi)$ such that $u := q^{-1}g \in \mathrm{G}$. Hence

$$p := u^{-1}g = g^{-1}qg \in \mathrm{P}(g^{-1}\xi g) = \mathrm{P}(\zeta)$$

and

$$p\zeta p^{-1} = u^{-1}g\zeta g^{-1}u = u^{-1}\xi u \in \mathfrak{g}.$$

This shows that every equivalence class in $\mathscr{T}^c$ contains an element of $\mathfrak{g}$.

We prove uniqueness. Let $\xi \in \mathfrak{g} \setminus \{0\}$ and $p \in \mathrm{P}(\xi)$ such that $p\xi p^{-1} \in \mathfrak{g}$. Choose the eigenvalues $\lambda_1 < \cdots < \lambda_k$ of $\mathbf{i}\xi$ and the eigenspace decomposition $\mathbb{C}^n = V_1 \oplus \cdots \oplus V_k$ as in the proof of Lemma D.2. Since $\mathbf{i}\xi$ is Hermitian its eigenspaces V_j are pairwise orthogonal. Moreover, the subspace $V_1 \oplus \cdots \oplus V_j$ is invariant under p and hence also under $p\xi p^{-1}$ for every j. Since $p\xi p^{-1} \in \mathfrak{g}$ is a skew-Hermitian endomorphism of $\mathbb{C}^n$ and the complex subspaces $V_1, \ldots, V_k$ of $\mathbb{C}^n$ are pairwise orthogonal, it follows that

$$p\xi p^{-1} V_i \subset V_i, \qquad i = 1, \ldots, k.$$

Hence $p\xi p^{-1} = \xi$. This completes the proof of Theorem D.4. □

Appendix E
The Partial Flag Manifold $\mathbf{G}^c/\mathbf{P} \equiv \mathbf{G}/\mathbf{C}$

In this appendix we prove Theorem D.3.

Lemma E.1 *Let $N \in \mathbb{N}$. There exist real numbers*

$$\beta_0(N), \beta_1(N), \ldots, \beta_{2N-1}(N)$$

such that $\beta_\nu(N) = 0$ when ν is even and, for $k = 1, 3, 5, \ldots, 4N-1$,

$$\sum_{\nu=0}^{2N-1} \beta_\nu(N) \exp\left(\frac{k\nu\pi\mathbf{i}}{2N}\right) = \begin{cases} \mathbf{i}, \ \textit{if}\ 0 < k < 2N, \\ -\mathbf{i}, \ \textit{if}\ 2N < k < 4N. \end{cases} \tag{E.1}$$

Proof Define

$$\lambda := \exp\left(\frac{\pi\mathbf{i}}{2N}\right)$$

and consider the Vandermonde matrix

$$\Lambda := \begin{pmatrix} \lambda & \lambda^3 & \lambda^5 & \ldots & \lambda^{2N-1} \\ \lambda^3 & \lambda^9 & \lambda^{15} & \ldots & \lambda^{6N-3} \\ \lambda^5 & \lambda^{15} & \lambda^{25} & \ldots & \lambda^{10N-5} \\ \vdots & \vdots & \vdots & \ddots & \vdots \\ \lambda^{2N-1} & \lambda^{6N-3} & \lambda^{10N-5} & \ldots & \lambda^{(2N-1)^2} \end{pmatrix} \in \mathbb{C}^{N\times N}.$$

Its complex determinant is

$$\det{}^c(\Lambda) = \lambda^{N(2N-1)} \prod_{0\le i<j\le N-1} \left(\lambda^{4j} - \lambda^{4i}\right).$$

© The Author(s), under exclusive license to Springer Nature Switzerland AG 2021

V. Georgoulas et al., *The Moment-Weight Inequality and the Hilbert–Mumford Criterion*, Lecture Notes in Mathematics 2297,
https://doi.org/10.1007/978-3-030-89300-2

Since λ is a primitive $4N$th root of unity, the numbers

$$\lambda^{4i}, \qquad i = 0, \ldots, N-1,$$

are pairwise distinct. Hence Λ is nonsingular.

Since Λ is nonsingular, there exists a unique vector

$$z = (z_1, z_3, \ldots, z_{2N-1}) \in \mathbb{C}^N$$

such that

$$\sum_{\substack{0<\nu<2N \\ \nu \text{ odd}}} \exp\left(\frac{k\nu\pi\mathbf{i}}{2N}\right) z_\nu = \mathbf{i}, \qquad k = 1, 3, \ldots, 2N-1. \tag{E.2}$$

The numbers $\overline{z}_\nu$ also satisfy Eq. (E.2), because

$$\sum_{\substack{0<\nu<2N \\ \nu \text{ odd}}} \exp\left(\frac{k\nu\pi\mathbf{i}}{2N}\right) \overline{z}_\nu = -\overline{\sum_{\substack{0<\nu<2N \\ \nu \text{ odd}}} \exp\left(\frac{(2N-k)\nu\pi\mathbf{i}}{2N}\right) z_\nu} = \mathbf{i}$$

for $k = 1, 3, \ldots, 2N-1$. Since the solution is unique the z_ν are real.

Now define the numbers

$$\beta_0(N), \beta_1(N), \ldots, \beta_{2N-1}(N)$$

by

$$\beta_\nu(N) := \begin{cases} z_\nu, & \text{for } \nu = 1, 3, \ldots, 2N-1, \\ 0, & \text{for } \nu \text{ even}. \end{cases}$$

These numbers satisfy (E.1) for $k = 1, 3, \ldots, 2N-1$ by (E.2). Equation (E.1) also holds for $k = 2N+1, 2N+3, \ldots, 4N-1$ because $\exp(k\pi\mathbf{i}) = -1$ whenever k is odd. This proves Lemma E.1. □

Lemma E.2 *Let $m \in \mathbb{N}$ and $N := 2^m$. There exist real numbers*

$$\alpha_0(N), \alpha_1(N), \ldots, \alpha_{2N-1}(N)$$

such that $\alpha_\nu(N) = 0$ when ν is even and, for every $k \in \{0, 1, \ldots, 2N-1\}$,

$$\sum_{\nu=0}^{2N-1} \alpha_\nu(N) \exp\left(\frac{k\nu\pi\mathbf{i}}{N}\right) = \begin{cases} \mathbf{i}, & \text{if } 1 \le k \le N-1, \\ -\mathbf{i}, & \text{if } N+1 \le k \le 2N-1, \\ 0, & \text{if } k = 0 \text{ or } k = N. \end{cases} \tag{E.3}$$

Proof The proof is by induction on m. For

$$m = 1, \qquad N = 2^m = 2$$

choose

$$\alpha_1(2) := \tfrac{1}{2}, \qquad \alpha_3(2) := -\tfrac{1}{2}.$$

Then

$$\sum_{\nu=0}^{3} \alpha_\nu(2) \exp\left(\frac{k\nu\pi\mathbf{i}}{2}\right) = \frac{\mathbf{i}^k - (-\mathbf{i})^k}{2} = \begin{cases} \mathbf{i}, \text{ for } k = 1, \\ -\mathbf{i}, \text{ for } k = 3, \\ 0, \text{ for } k = 0, 2. \end{cases}$$

Now let $m \in \mathbb{N}$ and define $N := 2^m$. Assume, by induction, that the numbers

$$\alpha_\nu(N), \qquad \nu = 0, 1, \ldots, 2N - 1,$$

have been found such that (E.3) holds for $k = 0, 1, \ldots, 2N - 1$. Let

$$\beta_\nu(N), \qquad \nu = 0, 1, \ldots, 2N - 1,$$

be the constants of Lemma E.1. Define

$$\alpha_{2N+\nu}(N) := \alpha_\nu(N), \qquad \beta_{2N+\nu}(N) := -\beta_\nu(N),$$

for $\nu = 0, 1, 2, \ldots, 2N - 1$ and

$$\alpha_\nu(2N) := \frac{\alpha_\nu(N) + \beta_\nu(N)}{2}, \qquad \nu = 0, 1, 2, \ldots, 4N - 1. \tag{E.4}$$

Then

$$\sum_{\nu=0}^{4N-1} \alpha_\nu(2N) \exp\left(\frac{k\nu\pi\mathbf{i}}{2N}\right) = A_k + B_k,$$

where

$$A_k := \frac{1}{2} \sum_{\nu=0}^{4N-1} \alpha_\nu(N) \exp\left(\frac{k\nu\pi\mathbf{i}}{2N}\right),$$

$$B_k := \frac{1}{2} \sum_{\nu=0}^{4N-1} \beta_\nu(N) \exp\left(\frac{k\nu\pi\mathbf{i}}{2N}\right).$$

Since $\alpha_{2N+\nu}(N) = \alpha_\nu(N)$, we have

$$A_k = \frac{1}{2}\sum_{\nu=0}^{4N-1} \alpha_\nu(N) \exp\left(\frac{k\nu\pi\mathbf{i}}{2N}\right)$$
$$= \frac{1+\exp(k\pi\mathbf{i})}{2}\sum_{\nu=0}^{2N-1} \alpha_\nu(N) \exp\left(\frac{k\nu\pi\mathbf{i}}{2N}\right)$$
$$= \frac{1+(-1)^k}{2}\sum_{\nu=0}^{2N-1} \alpha_\nu(N) \exp\left(\frac{k\nu\pi\mathbf{i}}{2N}\right).$$

If k is odd, then the right hand side vanishes. If k is even, then by the induction hypothesis,

$$A_k = \sum_{\nu=0}^{2N-1} \alpha_\nu(N) \exp\left(\frac{(k/2)\nu\pi\mathbf{i}}{N}\right)$$
$$= \begin{cases} \mathbf{i}, \text{ for } k = 2, 4, \ldots, 2N-2, \\ -\mathbf{i}, \text{ for } k = 2N+2, \ldots, 4N-2, \\ 0, \text{ for } k = 0, 2N. \end{cases}$$

Since $\beta_{2N+\nu}(N) = -\beta_\nu(N)$, we have

$$B_k = \frac{1}{2}\sum_{\nu=0}^{2N-1} \beta_\nu(N)\left(\exp\left(\frac{k\nu\pi\mathbf{i}}{2N}\right) - \exp\left(\frac{k(2N+\nu)\pi\mathbf{i}}{2N}\right)\right)$$
$$= \frac{1-\exp(k\pi\mathbf{i})}{2}\sum_{\nu=0}^{2N-1} \beta_\nu(N) \exp\left(\frac{k\nu\pi\mathbf{i}}{2N}\right)$$
$$= \frac{1-(-1)^k}{2}\sum_{\nu=0}^{2N-1} \beta_\nu(N) \exp\left(\frac{k\nu\pi\mathbf{i}}{2N}\right).$$

If k is even, then the right hand side vanishes. If k is odd, then by Lemma E.1,

$$B_k = \sum_{\nu=0}^{2N-1} \beta_\nu(N) \exp\left(\frac{k\nu\pi\mathbf{i}}{2N}\right)$$
$$= \begin{cases} \mathbf{i}, \text{ if } k = 1, 3, \ldots, 2N-1, \\ -\mathbf{i}, \text{ if } k = 2N+1, \ldots, 4N-1. \end{cases}$$

Combining the formulas for A_k and B_k we find

$$A_k + B_k = \begin{cases} \mathbf{i}, \text{ for } k = 1, 2, 3, \ldots, 2N-1, \\ -\mathbf{i}, \text{ for } k = 2N+1, 2N+2, \ldots, 4N-1, \\ 0, \text{ for } k = 0, 2N, \end{cases}$$

and this proves Lemma E.2. □

Lemma E.3 *Let* $\xi, \eta \in \mathfrak{g} \subset \mathfrak{su}(n)$ *and assume* $\exp(\xi) = \mathbb{1}$. *Then there exists an element* $\zeta \in \mathfrak{p}(\xi)$ *such that* $\zeta - \mathbf{i}\eta \in \mathfrak{g}$.

Proof Let $\lambda_1 < \cdots < \lambda_k$ be the eigenvalues of the Hermitian matrix $\mathbf{i}\xi$ and denote the corresponding eigenspace decomposition by

$$\mathbb{C}^n = V_1 \oplus \cdots \oplus V_k.$$

Then

$$\lambda_i - \lambda_j = 2\pi m_{ij}, \qquad m_{ij} \in \mathbb{Z},$$

with $m_{ij} > 0$ for $i > j$ and $m_{ij} < 0$ for $i < j$. Choose $m \in \mathbb{N}$ such that

$$N := 2^m > m_{k1} = \frac{\lambda_k - \lambda_1}{2\pi}.$$

Choose $\alpha_0, \ldots, \alpha_{2N-1} \in \mathbb{R}$ as in Lemma E.2. Let

$$\eta = \begin{pmatrix} \eta_{11} & \eta_{12} & \cdots & \eta_{1k} \\ \eta_{21} & \eta_{22} & \cdots & \eta_{2k} \\ \vdots & \vdots & \ddots & \vdots \\ \eta_{k1} & \eta_{k2} & \cdots & \eta_{kk} \end{pmatrix} \in \mathfrak{g}, \qquad \eta_{ij} \in \mathrm{Hom}(V_j, V_i).$$

Define

$$\zeta := \mathbf{i}\eta - \sum_{\nu=0}^{2N-1} \alpha_\nu \exp\left(-\frac{\nu}{2N}\xi\right) \eta \exp\left(\frac{\nu}{2N}\xi\right) \in \mathfrak{g}^c.$$

Then, for $i > j$, we have

$$\begin{aligned}\zeta_{ij} &= \mathbf{i}\eta_{ij} - \sum_{\nu=0}^{2N-1} \alpha_\nu \exp\left(\frac{\nu}{2N}\mathbf{i}(\lambda_i - \lambda_j)\right) \eta_{ij} \\ &= \left(\mathbf{i} - \sum_{\nu=0}^{2N-1} \alpha_\nu \exp\left(\frac{m_{ij}\nu\pi\mathbf{i}}{N}\right)\right) \eta_{ij} \\ &= 0.\end{aligned}$$

The last equation follows from Lemma E.2 and the fact that $1 \le m_{ij} \le N-1$ for $i > j$. Since $\zeta_{ij} = 0$ for $i > j$ it follows from the proof of Lemma D.2 that $\zeta \in \mathfrak{p}(\xi)$. Moreover, by construction $\mathbf{i}\eta - \zeta \in \mathfrak{g}$. This proves Lemma E.3. □

Proof of Theorem D.3 Assume first that $\xi \in \Lambda$ so that $\exp(\xi) = \mathbb{1}$. Define

$$A := \left\{ g \in G^c \;\middle|\; \begin{array}{l}\text{there exists an element } p \in P(\xi) \\ \text{such that } p^{-1}g \in G\end{array} \right\}.$$

We will prove by an open and closed argument that $A = G^c$.

We prove that A is a closed subset of G^c (in the relative topology). Let $g_i \in A$ be a sequence which converges to an element $g \in G^c$. Then there exists a sequence $p_i \in P(\xi)$ such that

$$u_i := p_i^{-1} g_i \in G.$$

Since G is compact there exists a subsequence (still denoted by u_i) which converges to an element $u \in G$. Since $P(\xi)$ is a closed subset of G^c, we have

$$p := gu^{-1} = \lim_{i\to\infty} g_i u_i^{-1} = \lim_{i\to\infty} p_i \in P(\xi).$$

Hence $p^{-1}g = u \in G$ and so $g \in A$. Thus A is a closed subset of G^c.

We prove that the map $f : P(\xi) \times G \to G^c$, defined by

$$f(p, u) := pu$$

for $p \in P(\xi)$ and $u \in G$, is a submersion. Let $p \in P(\xi)$ and $u \in G$ and denote

$$g := f(p, u) = pu.$$

Let $\widehat{g} \in T_g G^c$ and denote

$$\widetilde{\zeta} := p^{-1}\widehat{g}u^{-1} = u(g^{-1}\widehat{g})u^{-1} \in \mathfrak{g}^c. \tag{E.5}$$

Let $\eta \in \mathfrak{g}$ be the imginary part of $\widetilde{\zeta}$ so that $\widetilde{\zeta} - \mathbf{i}\eta \in \mathfrak{g}$. By Lemma E.3, there exists an element $\zeta \in \mathfrak{p}(\xi)$ such that $\zeta - \mathbf{i}\eta \in \mathfrak{g}$ and hence $\widetilde{\zeta} - \zeta \in \mathfrak{g}$. Define

$$\widehat{p} := p\zeta, \qquad \widehat{u} := \big(\widetilde{\zeta} - \zeta\big)u.$$

Then $\widehat{p} \in T_p\mathrm{P}(\xi)$, $\widehat{u} \in T_u\mathrm{G}$, and

$$df(p,u)(\widehat{p},\widehat{u}) = \widehat{p}u + p\widehat{u} = p\widetilde{\zeta}u = \widehat{g}.$$

Here the last equation follows from (E.5). Thus we have proved that the derivative $df(p,u) : T_p\mathrm{P}(\xi) \times T_u\mathrm{G} \to T_{pu}\mathrm{G}^c$ is surjective for every $p \in \mathrm{P}(\xi)$ and every $u \in \mathrm{G}$. Hence f is a submersion as claimed.

We prove that $A = \mathrm{G}^c$. The set A contains G by definition. Moreover, we have proved that it is closed and that it is the image of a submersion and hence is open. Since G^c is homeomorphic to $\mathrm{G} \times \mathfrak{g}$ and A contains $\mathrm{G} \cong \mathrm{G} \times \{0\}$, it follows that A intersects each connected component of G^c in a nonempty open and closed set. Hence $A = \mathrm{G}^c$. This proves Theorem D.3 for $\xi \in \Lambda$.

Now let $\xi \in \mathfrak{g}\setminus\{0\}$. Choose sequences $\xi_i \in \Lambda$ and $s_i \in \mathbb{R}$ such that $s_i\xi_i$ converges to ξ. By the first part of the proof there exist sequences $p_i \in \mathrm{P}(\xi_i)$ and $u_i \in \mathrm{G}$ such that $u_i p_i = g$ for every i. Passing to a subsequence if necessary we may assume that u_i converges to $u \in \mathrm{G}$. Hence $p_i = u_i^{-1}g$ converges to $p = u^{-1}g \in \mathrm{G}^c$. Examining the eigenspace decompositions of ξ and ξ_i we find that $p \in \mathrm{P}(\xi)$. This proves Theorem D.3. □

References

1. Michael Atiyah, Convexity and commuting Hamiltonians. *Bulletin of the London Mathematical Society* **14** (1982), 1–15.
2. Michael Atiyah & Raoul Bott, The Yang-Mills equations over Riemann surfaces. *Phil. Trans. Roy. Soc. London A* **308** (1982), 523–615.
3. M.F. Atiyah & V.G. Drinfeld & N.J. Hitchin & Y.I. Manin Construction of instantons. *Physics Letters A* **65** (1978), 185–187.
4. Robert J. Berman & Bo Berndtsson, Convexity of the K-energy on the space of Kahler metrics and uniqueness of extremal metrics. Preprint, 2 May 2014. https://arxiv.org/abs/1405.0401
5. Edward Bierstone & Pierre E. Milman, Semianalytic and subanalytic sets. *Publications Mathématiques de l'I.H.É.S.* **67** (1988), 5–42.
6. Eugenio Calabi, The space of Kähler metrics. *Proceedings of the International Congress of Mathematicians, Amsterdam* **2** (1954), 206–207.
7. Eugenio Calabi, On Kähler manifolds with vanishing canonical class, *Algebraic geometry and topology. A symposium in honor of S. Lefschetz*, edited by Ralph H. Fox, D.C. Spencer, A.W. Tucker, Princeton Mathematical Series **12**, Princeton University Press, 1957, pp 78–89.
8. Eugenio Calabi, Extremal Kähler metrics. *Seminar on differential geometry*, edited by Yau et al, *Annals of Math. Studies* **102**, Princeton University Press, 1982.
9. Eugenio Calabi & Xiuxiong Chen, Space of Kähler metrics and Calabi flow. *Journal of Differential Geometry* **61** (2002), 173–193.
10. Élie Joseph Cartan, Les groupes réels simples, finis et continus. *Annales scientifiques de l'École Normale Sup'erieure* **31** (1914), 263–355.
11. Élie Joseph Cartan, Groupes simples clos et ouverts et géométrie Riemannienne. *Journal de Mathématiques Pures et Appliquées* **8** (1929) 1–33.
12. Élie Joseph Cartan, La théorie des groupes finis et continus et l'Analysis Situs. *Mémorial des Sciences Mathématiques* **42** (1930), 1–61.
13. Bill Casselmann, Symmetric spaces of semi-simple groups. *Essays on representations of real groups*, August 2012. http://www.math.ubc.ca/~cass/research/pdf/Cartan.pdf
14. Xiuxiong Chen, Space of Kähler metrics. *Journal of Differential Geometry* **56** (2000), 189–234.
15. Xiuxiong Chen, Space of Kähler metrics III—On the lower bound of the Calabi energy and geodesic distance. *Inventiones Mathematicae* **175** (2009), 453–503.
16. Xiuxiong Chen, Space of Kähler metrics IV—On the lower bound of the K-energy. September 2008. http://arxiv.org/abs/0809.4081v2

© The Author(s), under exclusive license to Springer Nature Switzerland AG 2021

V. Georgoulas et al., *The Moment-Weight Inequality and the Hilbert–Mumford Criterion*, Lecture Notes in Mathematics 2297,
https://doi.org/10.1007/978-3-030-89300-2

17. Xiuxiong Chen & Simon Donaldson & Song Sun, Kähler-Einstein metrics and stability. *International Mathematics Research Notices* **2013**.
18. Xiuxiong Chen & Simon Donaldson & Song Sun, Kähler-Einstein metrics on Fano manifolds, I: approximation of metrics with cone singularities. *JAMS* **28** (2015), 183–197.
19. Xiuxiong Chen & Simon Donaldson & Song Sun, Kähler-Einstein metrics on Fano manifolds, II: limits with cone angle less than 2π. *JAMS* **28** (2015), 199–234.
20. Xiuxiong Chen & Simon Donaldson & Song Sun, Kähler-Einstein metrics on Fano manifolds, III: limits as cone angle approaches 2π and completion of the main proof. *JAMS* **28** (2015), 235–278.
21. Xiuxiong Chen & Long Li & Mihai Paun, Approximation of weak geodesics and subharmonicity of Mabuchi energy. Preprint, 28 September 2014. https://arxiv.org/abs/1409.7896
22. Xiuxiong Chen & Song Sun, Space of Kähler metrics V–Kähler quantization. April 2010. http://arxiv.org/abs/0902.4149v2
23. Xiuxiong Chen & Song Sun, Calabi flow, Geodesic rays, and uniqueness of constant scalar curvature Kähler metrics. *Annals of Mathematics* **180** (2014), 407–454.
24. Xiuxiong Chen & Song Sun & Bing Wang, Kähler–Ricci flow, Kähler–Einstein metric, and K-stability. Preprint, 19 August 2015. https://arxiv.org/abs/1508.04397v1
25. Xiuxiong Chen & Gang Tian, Geometry of Kähler metrics and foliations by holomorphic discs, *Publications Mathématiques de I.H.É.S.* **107** (2008), 1–107. http://arxiv.org/abs/math/0507148.pdf
26. Tamás Darvas, The Mabuchi completion of the space of Kähler potentials. *American Journal of Mathematics* **139** (2017), 1275–1313.
27. Simon Donaldson, A new proof of a theorem of Narasimhan and Seshadri. *Journal of Differential Geometry* **18** (1983), 269–277.
28. Simon Donaldson, Anti-self-dual Yang-Mills connections on complex algebraic surfaces and stable vector bundles. *Proc. London Math. Soc.* **50** (1985), 1–26.
29. Simon Donaldson, Moment Maps and Diffeomorphisms. *Asian Journal of Mathematics* **3** (1999), 1–16.
30. Simon Donaldson, Symmetric spaces, Kähler geometry and Hamiltonian dynamics. *Northern California Symplectic Geometry Seminar*, edited by Eliashberg et al, *Amer. Math. Soc. Transl. Ser.* 2, **196**, 1999, 13–33.
31. Simon Donaldson, Scalar curvature and projective embeddings, I. *Journal of Differential Geometry* **59** (2001), 479–522.
32. Simon Donaldson, Conjectures in Kähler geometry. *Strings and Geometry*, Proceedings of the Clay Mathematics Institute 2002 Summer School on Strings and Geometry, Isaac Newton Institute, Cambridge, United Kingdom, March 24–April 20, 2002. *Clay Mathematical Proceedings* **3**, AMS 2004, pp 71–78.
33. Simon Donaldson, Scalar curvature and stability of toric varieties. *Journal of Differential Geometry* **62** (2002), 289–349.
34. Simon Donaldson, Lower bounds on the Calabi functional. *Journal of Differential Geometry* **70** (2005), 453–472.
35. Simon Donaldson, Scalar curvature and projective embeddings, II. *Quarterly Journal of Mathematics* **56** (2005), 345–356.
36. Simon Donaldson, Lie algebra theory without algebra. In *"Algebra, Arithmetic, and Geometry, In honour of Yu. I. Manin"*, edited by Yuri Tschinkel and Yuri Zarhin, Progress in Mathematics, Birkhäuser **269**, 2009, pp 249–266.
37. Simon Donaldson & Song Sun, Gromov–Hausdorff limits of Kähler manifolds and algebraic geometry. Preprint, 12 June 2012. http://arxiv.org/abs/1206.2609
38. Yasha Eliashberg & Leonid Polterovich, Bi-invariant metrics on the group of Hamiltonian diffeomorphisms. *International Journal of Mathematics* **4** (1993), 727–738.
39. Akira Fujiki, Moduli spaces of polarized algebraic varieties and Kähler metrics. *Sûgaku* **42** (1990), 231–243. English Translation: *Sûgaku Expositions* **5** (1992), 173–191.
40. Akito Futaki, An obstruction to the existence of Einstein-Kähler metrics. *Inventiones Mathematicae* **73** (1983), 437–443.

41. Valentina Georgoulas, A differential geometric approach to GIT and stability. PhD thesis, ETH Zürich, 2016.
42. Victor Guillemin & Shlomo Sternberg, Geometric quantization and multiplicities of group represnetations. *Inventiones Mathamticae* **67** (1982), 515–538.
43. Alain Haraux, Some applications of the Lojasiewicz gradient inequality. *Communications on Pure and Applied Analysis* **6** (2012), 2417–2427.
44. Sigurour Helgason, *Differential Geometry, Lie Groups, and Symmetric Spaces*. Graduate Studies in Mathematics **35**, AMS, 2001.
45. Uwe Helmke, The topology of a moduli space for linear dynamical systems, *Commentarii Mathematici Helvetici* **60** (1985), 630–655.
46. Uwe Helmke, Topology of the moduli space for reachable linear dynamical systems: The complex case. *Mathematical Systems Theory* **19** (1986), 155–187.
47. Uwe Helmke, Linear dynamical systems and instantons in Yang-Mills theory, *IMA Journal of Mathematical Control and Information* **3** (1986), 151–166.
48. Uwe Helmke & Diederich Hinrichsen, Canonical Forms and Orbit Spaces of Linear Systems. *IMA Journal of Mathematical Control and Information* **3** (1986), 167–184.
49. Uwe Helmke & John B. Moore, *Opimization and Dynamical Systems*. Springer-Verlag, London, UK, 1994.
50. Joachim Hilgert & Karl-Hermann Neeb, *Structure and Geometry of Lie groups*. Springer Monographs in Mathematics, Springer-Verlag, 2012.
51. George Kempf, Instability in invariant theory. *Annals of Mathematics* **108** (1978), 299–317.
52. George Kempf & Linda Ness, The length of vectors in representation spaces. Springer Lecture Notes **732**, *Algebraic Geometry, Proceedings*, Copenhagen, 1978, pp 233–244.
53. Frances Kirwan, *Cohomology of Quotients in Symplectic and Algebraic Geometry*. Princeton University Press, 1984.
54. Peter B. Kronheimer, Simon Donaldson's Mathematics: A Retrospective. Lecture at the SCGP workshop *Geometry of Manifolds* on 26 October 2017. http://scgp.stonybrook.edu/video/video.php?id=3397
55. Jorge Lauret, On the moment map for the variety of Lie algebras. *Journal of Functional Analysis* **202** (2003), 392–423.
56. Eugene Lerman, Gradient flow of the norm squared of the moment map. *L'Enseignement Mathémathique* **51** (2005), 117–127.
57. Stanislaw Lojasiewicz, Une propriété topologique des sous-ensembles analytiques réels. *Les Équations aux Dérivées Partielles*, Éditions du Centre National de la Recherche Scientifique, Paris, *Colloques Internationaux du C.N.R.S.* **117** (1963), 87–89.
58. Toshiki Mabuchi, K-energy maps integrating Futaki invariants. *Tohoku Math. Journal* **38** (1986), 575–593.
59. Toshiki Mabuchi, Einstein–Kähler forms, Futaki invariants and convex geometry on toric Fano varieties. *Osaka Journal of Mathematics* **24** (1987), 705–737.
60. A.I. Malcev, On the theory of the Lie groups in the large. *Rec. Math. (Matematichevskii Sbornik)* **16(58)** (1945), 163–190.
61. Dusa McDuff & Dietmar Salamon, *J-holomorphic Curves and Symplectic Topology, Second Edition*. AMS Colloquium Publication **52**, 2012.
62. John Morgan & Tomasz Mrowka & Daniel Ruberman, *L^2 Moduli Spaces and a Vanishing Theorem for Donaldson Polynomial Invariants*. Monographs in Geometry and Topology, International Press, 1994.
63. David Mumford & John Fogarty & Frances Kirwan, *Geometric Invariant Theory, Third Enlarged Edition*. Ergebnisse der Mathematik und ihrer Grenzgebiete **34**, Springer Verlag, New York, 1994.
64. Mudumbai S. Narasimhan & Conjeevaram S. Seshadri, Stable and unitary vector bundles on compact Riemann surfaces. *Annals of Mathematics* **82** (1965), 540–567.
65. Linda Ness, Mumford's numerical function and stable projective hypersurfaces. Springer Lecture Notes **732**, *Algebraic Geometry, Proceedings*, Copenhagen, 1978, pp 417–454.

66. Linda Ness, A stratification of the null cone by the moment map. *American Journal of Mathematics* **106** (1984), 1281–1329.
67. Sean T. Paul, Hyperdiscriminant polytopes, Chow polytopes, and Mabuchi energy asymptotics. *Annals of Mathematics* **175** (2012), 255–296.
68. Sean T. Paul, Stable pairs and coercive estimates for the Mabuchi functional. Preprint, 20 August 2013. http://arxiv.org/abs/1308.4377
69. Roger Richardson, Compact real forms of a complex semisimple Lie algebra. *Journal of Differential Geometry* **2** (1968), 411–419.
70. Joel Robbin & Dietmar Salamon, *Introduction to Differential Geometry*, 2013. http://www.math.ethz.ch/~salamon/PREPRINTS/diffgeo.pdf
71. Leon Simon, Asymptotics for a class of nonlinear evolution equations, with applications to geometric problems. *Annals of Mathematics* **118** (1983), 525–571.
72. Gábor Székelyhidi, Extremal metrics and K-stability. PhD thesis, Imperial College, London, 2006. http://arxiv.org/pdf/math.DG/0611002.pdf
73. Gábor Szekélyhidi, The partial C^0 estimate along the continuity method. Preprint, 31 October 2013. http://arxiv.org/abs/1310.8471v1
74. Richard P. Thomas, Notes on GIT and symplectic reduction for bundles and varieties. *Surveys in Differential Geometry* **10**, *A tribute to S.-S. Chern*, edited by Shing-Tung Yau, Intern. Press, 2006. http://arxiv.org/abs/math.AG/0512411
75. Gang Tian, Kähler-Einstein metrics with positive scalar curvature. *Inventiones Mathematicae* **130** (1997), 1–37.
76. Karen Uhlenbeck & Shing-Tung Yau, On the existence of Hermitian Yang-Mills connections in stable vector bundles. *Communications on Pure and Applied Mathematics* **39** (1986), 257–293.
77. Lijing Wang, Hessians of the Calabi Functional and the Norm Function. *Annals of Global Analysis and Geometry* **29** (2006), 187–196.
78. Christopher T. Woodward, Moment maps and geometric invariant theory. June 2011. http://arxiv.org/abs/0912.1132v6
79. Shing-Tung Yau, Calabi's conjecture and some new results in algebraic geometry. *Proceedings of the National Academy of Sciences of the United States of America* **74** (1977), 1798–1799.
80. Shing-Tung Yau, On the Ricci curvature of a compact Kähler manifold and the complex Monge- Ampère equation I. *Communications in Pure and Applied Mathematics* **31** (1978), 339–411.
81. Shing-Tung Yau, Open problems in geometry. *L'Enseignement Mathémathique* **33** (1987), 109–158.

Index

© The Author(s), under exclusive license to Springer Nature Switzerland AG 2021

V. Georgoulas et al., *The Moment-Weight Inequality and the Hilbert–Mumford Criterion*, Lecture Notes in Mathematics 2297,
https://doi.org/10.1007/978-3-030-89300-2

LECTURE NOTES IN MATHEMATICS Springer

Editors in Chief: J.-M. Morel, B. Teissier;

Editorial Policy

1. Lecture Notes aim to report new developments in all areas of mathematics and their applications – quickly, informally and at a high level. Mathematical texts analysing new developments in modelling and numerical simulation are welcome.

   Manuscripts should be reasonably self-contained and rounded off. Thus they may, and often will, present not only results of the author but also related work by other people. They may be based on specialised lecture courses. Furthermore, the manuscripts should provide sufficient motivation, examples and applications. This clearly distinguishes Lecture Notes from journal articles or technical reports which normally are very concise. Articles intended for a journal but too long to be accepted by most journals, usually do not have this "lecture notes" character. For similar reasons it is unusual for doctoral theses to be accepted for the Lecture Notes series, though habilitation theses may be appropriate.

2. Besides monographs, multi-author manuscripts resulting from SUMMER SCHOOLS or similar INTENSIVE COURSES are welcome, provided their objective was held to present an active mathematical topic to an audience at the beginning or intermediate graduate level (a list of participants should be provided).

   The resulting manuscript should not be just a collection of course notes, but should require advance planning and coordination among the main lecturers. The subject matter should dictate the structure of the book. This structure should be motivated and explained in a scientific introduction, and the notation, references, index and formulation of results should be, if possible, unified by the editors. Each contribution should have an abstract and an introduction referring to the other contributions. In other words, more preparatory work must go into a multi-authored volume than simply assembling a disparate collection of papers, communicated at the event.

3. Manuscripts should be submitted either online at www.editorialmanager.com/lnm to Springer's mathematics editorial in Heidelberg, or electronically to one of the series editors. Authors should be aware that incomplete or insufficiently close-to-final manuscripts almost always result in longer refereeing times and nevertheless unclear referees' recommendations, making further refereeing of a final draft necessary. The strict minimum amount of material that will be considered should include a detailed outline describing the planned contents of each chapter, a bibliography and several sample chapters. Parallel submission of a manuscript to another publisher while under consideration for LNM is not acceptable and can lead to rejection.

4. In general, **monographs** will be sent out to at least 2 external referees for evaluation.

   A final decision to publish can be made only on the basis of the complete manuscript, however a refereeing process leading to a preliminary decision can be based on a pre-final or incomplete manuscript.

   Volume Editors of **multi-author works** are expected to arrange for the refereeing, to the usual scientific standards, of the individual contributions. If the resulting reports can be forwarded to the LNM Editorial Board, this is very helpful. If no reports are forwarded or if other questions remain unclear in respect of homogeneity etc, the series editors may wish to consult external referees for an overall evaluation of the volume.

5. Manuscripts should in general be submitted in English. Final manuscripts should contain at least 100 pages of mathematical text and should always include

   - a table of contents;
   - an informative introduction, with adequate motivation and perhaps some historical remarks: it should be accessible to a reader not intimately familiar with the topic treated;
   - a subject index: as a rule this is genuinely helpful for the reader.
   - For evaluation purposes, manuscripts should be submitted as pdf files.

6. Careful preparation of the manuscripts will help keep production time short besides ensuring satisfactory appearance of the finished book in print and online. After acceptance of the manuscript authors will be asked to prepare the final LaTeX source files (see LaTeX templates online: https://www.springer.com/gb/authors-editors/book-authors-editors/manuscriptpreparation/5636) plus the corresponding pdf- or zipped ps-file. The LaTeX source files are essential for producing the full-text online version of the book, see http://link.springer.com/bookseries/304 for the existing online volumes of LNM). The technical production of a Lecture Notes volume takes approximately 12 weeks. Additional instructions, if necessary, are available on request from lnm@springer.com.

7. Authors receive a total of 30 free copies of their volume and free access to their book on SpringerLink, but no royalties. They are entitled to a discount of 33.3 % on the price of Springer books purchased for their personal use, if ordering directly from Springer.

8. Commitment to publish is made by a *Publishing Agreement*; contributing authors of multiauthor books are requested to sign a *Consent to Publish form*. Springer-Verlag registers the copyright for each volume. Authors are free to reuse material contained in their LNM volumes in later publications: a brief written (or e-mail) request for formal permission is sufficient.

Addresses:

Professor Jean-Michel Morel, CMLA, École Normale Supérieure de Cachan, France
E-mail: moreljeanmichel@gmail.com

Professor Bernard Teissier, Equipe Géométrie et Dynamique,
Institut de Mathématiques de Jussieu – Paris Rive Gauche, Paris, France
E-mail: bernard.teissier@imj-prg.fr

Springer: Ute McCrory, Mathematics, Heidelberg, Germany,
E-mail: lnm@springer.com

Printed in the United States
by Baker & Taylor Publisher Services